Beer

Beer

3rd Edition

TAP INTO THE ART AND SCIENCE OF BREWING

CHARLES BAMFORTH

OXFORD
UNIVERSITY PRESS

2009

OXFORD
UNIVERSITY PRESS

Oxford University Press, Inc., publishes works that further
Oxford University's objective of excellence
in research, scholarship, and education.

Oxford New York
Auckland Cape Town Dar es Salaam Hong Kong Karachi
Kuala Lumpur Madrid Melbourne Mexico City Nairobi
New Delhi Shanghai Taipei Toronto

With offices in
Argentina Austria Brazil Chile Czech Republic France Greece
Guatemala Hungary Italy Japan Poland Portugal Singapore
South Korea Switzerland Thailand Turkey Ukraine Vietnam

Copyright © 2009 by Oxford University Press, Inc.

Published by Oxford University Press, Inc.
198 Madison Avenue, New York, NY 10016

www.oup.com

Oxford is a registered trademark of Oxford University Press

Library of Congress Cataloging-in-Publication Data
Bamforth, Charles W., 1952–
Beer : tap into the art and science of brewing / Charles Bamforth.—3rd ed.
p. cm.
Includes bibliographical references and index.
ISBN 978-0-19-530542-5
1. Brewing—Amateurs' manuals. 2. Beer.
I. Title.
TP577.B34 2009
641.6'23—dc22   2008050014

20 19 18 17 16 15 14 13 12 11

Printed in the United States of America
on acid-free paper

For Diane, Peter, Caroline, Emily, Stephanie, and Arek

Beer...a high and mighty liquor

JULIUS CAESAR

FOREWORD

The first edition of *Beer: Tap into the Art and Science of Brewing* was published in 1998. It is a great compliment to the quality and readability of this book that the third edition is being published only ten years later. There are a number (some would say too many!) of books and monographs that consider the scientific and technical aspects of brewing. Nevertheless, it is a reflection of this book's high standard that its publishers regard its republication as warranted. In his Foreword to the first edition Dave Thomas wrote: "Charlie has managed to elegantly combine many of the most fascinating and memorable points that arise from the study of beer and brewing within the pages of this text, making it an enjoyable read." This is also true of the third edition.

Why is this the case? It is not a detailed technical volume with too much scientific information that will collect dust on a bookshelf. Written in the author's characteristic style, it is a readable, comprehensive publication that considers the brewing process, relevant historical aspects, beer types old and new, new materials, relevant analytical procedures, and lastly (but not least) a crystal ball examination considering malting and brewing in the near and long term.

What are the strengths of this book? Firstly, as already discussed, it is easily readable, with many interesting anecdotes and historical details. Secondly, the photographs and other illustrations employed have been well researched and are of good quality. Thirdly, every effort has been made to include novel material; it is a real third edition, not just a reprinting. In my comments printed on the cover of the first edition I remarked "This book is highly recommended to anyone—old and young, newcomer and old hand in the industry—who wishes to broaden his or her knowledge of brewing." This statement is more appropriate today than it was ten years ago because many more people are embarking on a career in the craft brewing industry with minimal technical background of the process. For example, any novice in the industry will find the glossary and the description of scientific principles contained in this book invaluable.

If I have one questionable comment, it is with the title of this volume, which as it was published ten years ago is a bit late! Nevertheless, I have never believed that brewing ever was or ever will be an art. It was a craft that through the application of scientific and engineering principles became a technology!

GRAHAM G. STEWART
Emeritus Professor of Brewing
Heriot-Watt University, Edinburgh
Scotland, U.K.

PREFACE

I forget who said: "I am not paranoid but that doesn't mean that they aren't out to get me." It would be very easy to become suspicious, being the brewing professor in a state where, among other amazing locales, the beautiful Napa Valley and the even more beautiful Sonoma Valley are meccas for wine aficionados. And so when I declare my role in life, there are plenty of folks who look at me initially with curiosity and amusement and, struggling to say something relevant, offer: "Oh, hops!" If I were to ask them "which of these wonderful drinks, beer and wine, do you think is the more complex, more demanding to produce, fitted for accompanying a broader range of foods, and healthier for you, and all this usually at a far more realistic price?" they would certainly not get the right answer. For them wine is liquid gold, whereas beer is molten lead. Nothing could be further from the truth.

Of course, the other thing those winebibbers are forever saying to me is "it takes a lot of good beer to make good wine." If only they would realize how patronizing and offensive that saying is. The reality is that the better winemakers have learned a great deal from studying the developments pioneered in the brewing industry.

This is not a book in which I wish to dwell overtly on the comparisons between wine and beer. I perceive them both to be fantastic elements of a fabulous lifestyle. True, I think wine has often been pedestalized in a ludicrously exaggerated fashion, and I fear that many a vinous emperor is in danger of being seen for the clothes that they are really wearing. By contrast, though, beer has too often been taken for granted: a Ford Festiva to wine's Rolls Royce. It is time for the layperson and for those nontechnical folks associated with the beer industry to learn that it is actually brewing that is the leader scientifically and technically, truly the giant in every respect within the alcoholic beverages industry. I hope this book helps in that educational process.

PREFACE

Not long before leaving England to take up my role as Professor of Malting and Brewing Sciences at the University of California, Davis, in February 1999 I was the guest on a local radio show in Guildford. Two questions I remember well.

The first was, "Charlie, did you think as a young boy at school in Lancashire that you would one day be the beer professor in California?" to which the instinctive reply, of course, was, "Well, it'll be a lousy job but I guess somebody has to do it. It must be my debt to society." I was, of course, using irony, lest anybody think I meant it!

The second question was more irritating. "Charlie, how will you possibly be able to enjoy those weak and tasteless beers over there after so long drinking our lovely English ales?" I was composed, replying thus:

> It's horses for courses. If I am in a 300-year-old thatched West Sussex pub, my bald head scraping the ceiling, snow outside, a roaring log fire within, a plate of Shepherd's Pie to devour from atop a well-scrubbed oaken table of great antiquity, then a pint of flat, generously hopped ale is a delight. However, if I'm in a baseball stadium, seventh inning stretching with a pile of nachos topped with jalapeños and 40 degrees of Mr. Celsius' best frying my few follicles, then an ice-cold Bud is to die for. And, by the way, if you're talking "weak," then do remember that a U.S. lager will typically contain 20% more alcohol (at least) than an ale from England.

On that occasion they didn't ask me the usual question beamed at a brewing professor. "But what is your favorite beer?" Usually I reply, "One that's wet and alcoholic."

This is, of course, something I don't believe. Just like there are good and bad footballers, and good and bad vicars, indeed good and bad virtually anything, then there are certainly beers (rather too many of them) that are plainly deplorable. Unquestionably, though, the great brewers of the world invariably delight the customer with their wares. A great many gently flavored lagers are superb, and wonderfully consistent. They have to be, for they are unforgiving and will reveal any conceivable shortcoming in raw material, process, or packaging. Equally, I can take you to some intensely flavored ales that are completely out of balance and devoid of all drinkability. There is no simple correlation between excellence and depth and complexity of flavor.

Which is why I get hopelessly infuriated with self-styled beer gurus who pontificate about what an ale or a lager should or should not be and about what should and what should not be the raw materials and processes that ought to be used, without the remotest understanding of the real science and technology of the brewer's art and the trials and tribulations of everyday existence in a brewing company.

This book attempts to give a reasoned view on such issues from the perspective of a longtime brewing scientist, research manager, quality assurance manager, customer, and, latterly, the bloke with the best job in the world.

PREFACE

TO THE FIRST EDITION

A year or two ago I was idly flicking between television channels when I chanced upon a couple of people sipping beer and discussing their findings. One of these people has established a reputation as being something of a wine connoisseur and would appear to take particular pride in pinpointing the exact vintage of the bottle and the winery in which it was produced. For all I know, that person may be able to name the peasant who had trod the grapes and predicted their shoe size. With rather more certainty, however, I was able to conclude that their knowledge of beer was mediocre, or worse.

From time to time, too, I come across articles in the general press, which pontificate about beer in a manner not dissimilar to the aforementioned wine buff. I applaud the efforts of some of these authors for the manner by which they help maintain beer in the collective consciousness. I deplore it, however, when they attempt to pontificate on the rights and wrongs of brewing practice. It is galling when they dress up the taste and aroma of beer in ridiculous terminology. Personally, I have enormous difficulty reconciling the language they use with the tastes of the myriad of beers that I have had the great good fortune to consume across the world.

An analogous situation exists in my own "other life." While it is research into the science of brewing and beer that pays my mortgage and puts food in the mouths of my children, my hobby is to write articles about soccer. I hope (and believe) that they help contribute to the pleasure of the fans who read them, but I hope I would never be accused of trying to tell the professionals within the game of how to do their jobs. I might fairly articulate the views of an "outsider"—the fan's-eye view, but I trust that it's the coaches and the players within soccer that know their specialization and can deliver a product that will thrill and delight me.

Rather more is written about beer in the nonspecialist press by "fans" than by "professionals." There is room for both—and that is why I decided to write this book, in an attempt to partly redress the balance. In it, I have attempted to capture the proud history of brewing, which stretches back to a time when articles on the merits of beer will have been written on papyrus or scrawled in hieroglyphics on walls of clay. I have attempted to convey the somewhat complex science of brewing in straightforward terms, with particular emphasis on why the properties of beer are as they are. I have endeavored to show what are the sensible and meaningful ways in which beer quality can be described. And I have tried to entertain, without trivializing a proud and distinguished profession.

I like beer, and, like the majority of people working in the brewing industry, I care about it and about the people who drink it. In this book I draw attention to a myriad of recent studies that suggest that beer and other alcoholic drinks are beneficial components of the adult diet, provided they are consumed in moderation. I certainly have no intention of encouraging the irresponsible to abuse the pleasure that comes from drinking beer in moderation, at the right time, in the right place.

I want people to understand and appreciate their beer, and gain an insight into the devoted labors of all those whose combined efforts bring it to the glass: the farmer who grows the best barley; the hop grower cultivating a unique crop; the maltster, who converts barley into delicious malt; the brewer who combines malt and hop to feed a yeast that he or she will have protected for perhaps hundreds of years; the bartender who keeps the beer in top condition.

This book is about facts. Where there is scope for expressing opinions, then these are my own and not everyone in the brewing trade will necessarily agree with them. They have, though, been arrived in a career in the brewing profession approaching 25 years. From reading this book I hope you will help form a considered opinion about brewing and about beer—and become rather better acquainted with its art and science.

ACKNOWLEDGMENTS

People often say to me: how do you find time to write your books? I tell them "it's thanks to Charlie's Angels." I will explain. Many years ago there was a television program called *Charlie's Angels*. I never did watch it, although I believe it had something to do with a bevy of female beauties doing detective work in support of a bloke called Charlie. I, too, have a group of women close to me whose contributions to my life make it possible for me to indulge myself in doing the things that I love.

Top of the list, of course, is my beautiful, talented, and infinitely patient wife, Diane. She has been in my life for some 37 years and is the heart and pulse of the Bamforth family. She drinks very little beer and will confess that her eyes glaze over whenever I get too intricate with my explanations about the ins and outs of malting and brewing. But she has supported me in everything I have tried to do. Without Diane there would be no point.

We have been blessed with three wonderful children. Now, of course I know that angels can be male—and Peter is a fantastic guy in every way. But our Englishness makes intimacy between us a firm handshake. We are great mates, but in the spirit of the theme he knows why it is his lovely wife Stephanie that I include in my list of Charlie's Angels. As I write, Peter and Stephanie are expecting our first grandchild, and Diane and I are thrilled. Such are the important things in life.

Caroline and Emily are our stunning daughters: they get their looks from their mother (and intelligence, too, for Diane was so smart to marry me on October 9, 1976!). Son-in-law Arek is, like Peter, a much-loved honorary Angel!

My university existence is blessed by a number of angels, of which five are especially important to me. The departmental manager Karen Nofziger is the most remarkable administrator I have ever met. She is tireless, organized, superefficient—and a hoot into the bargain. And backing her up and looking after the minutiae of keeping the show on the road are the fantastically efficient Jamie Ruffolo (who likes all things English) and Jamie Brannan. Without any of them I would be in chaos. And there is Melissa Haworth, my development guru, whose company I have so much enjoyed as we have wandered up and down California's immense Central Valley searching for the dollars to build new facilities.

In the laboratory and pilot brewery there is one of the most loyal and dedicated folks that it has been my pleasure to work with on the technical side of brewing. It's Candy Wallin who shoulders the burden of the critical but sometimes mundane tasks that go into keeping teaching, research, and pilot laboratories functioning. She does this with aplomb, devotion, and passion.

When my students see me bouncing around the lecture hall, gushing about beer and brewing, they can have no idea that the real me is actually a somewhat private individual, never happier than when sitting in his own "patchouli corner" in quiet meditation. It is those moments that fire my spirits and energize me for all the other things that I expect (and am expected) to do. For introducing me to, and continuing to guide me in, matters spiritual I will be eternally grateful to the incomparable Donna Stevens.

Apart from these wonderful people, I am so grateful to my patient publisher, Jeremy Lewis, who was assured receipt of the manuscript a full two years before he actually took hold of it, and also to Larry Nelson, editor of the *Brewer's Guardian*, for allowing me to draw on essays that I have written for his excellent publication over the years. Last, but never least, thanks to my biggest buddy in the brewing industry, Graham Stewart, for writing the foreword to this edition.

CONTENTS

INTRODUCTION

Ralph Waldo Emerson (1803–1882), the great American essayist, poet, and one-time Unitarian minister, penned many learned thoughts. The reader will forgive me if I select just thirteen words from the great man: "God made yeast…and loves fermentation just as dearly as he loves vegetation." Beer, surely, is a gift of God, one that brings together yeast and vegetation (in the shape of barley and hops) in a drink that has been enjoyed for 8,000 years, a beverage that has soothed fevered brows, nourished the hungry, coupled friendly and unfriendly alike—it's even seen men off into battle. "No soldier can fight unless he is properly fed on beef and beer," said John Churchill, the first Duke of Marlborough (1650–1722), a great British tactician and a forebear of the even more celebrated Winston.

Queen Victoria (1819–1901) was another who recognized the merit of beer: "Give my people plenty of beer, good beer and cheap beer, and you will have no revolution among them." With these sentiments, the redoubtable monarch echoed the enthusiasm of the Athenian tragedian Euripides (484–406 BC):

The man that isn't jolly after drinking
Is just a driveling idiot, to my thinking.

This book is not an exercise in trying to convince you, the reader, of the merits and demerits of drinking beer. I assume that as you have picked it up, and are starting to read, that you have an existing interest in beer. The aim of this book is to bring to the nonspecialist a feel for the beautiful complexity that underpins a truly international beverage. I use the word "international," accepting that to do justice to the entire world of beer would have demanded more than a single volume. Markets differ considerably from country to country, but the principles of brewing are constant the world over.

I have several audiences in mind for this book. First, and perhaps foremost, are the laymen who want to know, in reader-friendly terms, what goes into their beer. Such people seek the magic that underpins this supreme alchemy, namely the conversion of barley and hops by yeast into something so astonishingly drinkable. My desire is to reinforce the pleasure that people have in responsibly drinking beer by informing them, in a way hopefully accessible to even the least scientific among us, of what is involved in the production of their favorite drink and show them why malting and brewing are two of the great "traditional" industries. It is my earnest hope that, by reading this book, beer drinkers will appreciate the care that goes into making every pint of beer. I will be describing a science, or rather a range of sciences, and so can't avoid using scientific terms. Hopefully I have achieved this in a way readily understandable by those without an understanding of chemistry, biology, physics, chemical engineering, and the other scientific disciplines upon which brewing is founded. I have provided a simple explanation of the underpinning science and also a comprehensive glossary at the end of the book.

A second group of people who should benefit from reading this book are those joining the brewing industry who wish to have a "friendly" introduction to man's oldest biotechnology. Among these readers will be those entering in nontechnical roles: sales, marketing, finance...chief executives!

Thirdly there are those who interact professionally with brewing, either as suppliers or retailers, and need to know why the brewer is so demanding in his or her requirements and is so very proud of their heritage.

A valued colleague has extremely strong views on the use of language in books and lectures about the brewing business. I well recall having finished giving a lecture in Canada that I thought had gone across very nicely, when she stormed up to me, asking darkly whether I had a view on whether women as well as men drank beer. Puzzled, I replied that "of course they do." "Then why," she replied, acidly, "did every reference to the beer drinker in your talk consist of "he this," "he that." I had meant no offense by it, using "he" in the generic sense, but I haven't made the mistake since. For this reason, I intersperse the words "he" and "she" throughout this book. I haven't counted which term is used with greater frequency. As we will see in Chapter One, it is the female of the species that was once primarily responsible for brewing the ale. She was the "brewster," and I use that term, too, from time to time.

Another problem I had to confront was the matter of units. Brewers in different countries have their own scales of measurement. Even when the same name is used for a unit, it doesn't necessarily mean the same thing in different countries. Thus, a barrel in the United States comprises 31 gallons, whereas a barrel in the United Kingdom holds 36 gallons—and just to complicate matters further, a U.S. gallon is smaller than a U.K. gallon. I have used both types of "barrel" at various points and have indicated whether it's a U.S. or a U.K. variant. The international unit for volume, however, is the liter or the hectoliter (100 liters). By and large, this and other metric terms are employed, because brewers across the world do tend to use them as well as their own parochial preferences. A gallon equates to 3.7853 liters in the United States, hence a U.S. barrel holds 1.1734 hectoliters (hl). A U.K. barrel, on the other hand, contains 1.6365 hl because 1 U.K. gallon equals 1.201 U.S. gallons.

When I talk about the alcoholic strength of beer it is always as % vol./vol., which many people refer to as "alcohol by volume" (ABV). Thus 5% ABV indicates that there are 5ml (cm³) of alcohol (ethanol, previously known as ethyl alcohol) per 100 ml of beer. Although brewers in the United States still frequently use the Fahrenheit scale (and even until relatively recently the Reaumur scale, in which pure water freezes at 0 just like in the Celsius scale, but where its boiling point is at 80°), the rest of the brewing world uses degrees Celsius. In this edition I have softened, and given the reader both. Finally in connection with units, from time to time I talk about the levels of other molecules in beer, especially the substances that contribute to flavor. By and large these are present at quite low concentrations. You will find mention of ppm, ppb, and ppt: these refer to parts per million, parts per billion, and parts per trillion, respectively. A substance present at 1 ppm exists as 1 mg (milligram) per liter of beer. One ppb equates to 1 μg (microgram) per liter of beer, while 1 ppt means 1 ng (nanogram) per liter of beer. One mg is a thousandth of a gram; one microgram is a thousandth of a milligram; one nanogram is a thousandth of a microgram. Just in case the metric system still leaves you cold, I had better point out that there are 28.35 grams per ounce and 28 fluid ounces per U.S. gallon.

I also had the thorny question of which currency to use. As this book emanates from a New York publisher, I have chosen to use dollars and cents. Finally, when presenting statistical data, I have used the most recent information available to me. I regret that much of the trend information takes rather a long time for researchers to collect, so some details are a year or two old now.

Finally, a brief word about convention. I use Brewer (or Maltster) with a capital letter when referring to a brewing (or malting) *company*, but brewer and maltster in the lower case when describing an individual practitioner of the art.

Enjoy the book—and savor the beer that is the end result of so much care and devotion.

Beer

FROM SUMERIA TO SAN FRANCISCO

THE WORLD OF BEER AND BREWERIES

THE WORLD BEER MARKET

Beer is drunk all over the world. In some places, such as parts of Germany, it is *the* drink of choice for accompanying food. I well remember sitting in a restaurant near Munich witnessing the arrival of a coach loaded with elderly ladies and being astonished to see them demolish liters of lager, whereas the grannies I had been used to in England sipped tea. Across the globe, beer is the great drink of relaxation—and moderation. It is consumed in bars, clubs, and sports grounds—in fact anywhere where adults congregate. Surely nowhere typifies this better than the English public house (or "pub"), which remains, alongside the church and cricket wicket, as the essential ingredient of any self-respecting community, albeit changed insofar as food rather than beer seems now to be the prime magnet to many hostelries.

Yet it is clearly not essential to have company to pursue one's favorite tipple. In the United Kingdom, for instance, the proportion of beer sold on draft has declined from 78.3% when I first joined the industry in 1978 to 54.7% in 2005. Furthermore, in that time the proportion of beer sold in nonreturnable bottles (NRBs) has leaped from 0.5% to 14.6% of all beer sold. In the United States of 1933, 68% of all beer sold was "on tap," but by 2004 that had declined to 9%. There has been a clear shift toward beer drinking at home, driven in part by the increasing trend toward seeking one's social pleasures through entertaining or in more solitary pursuits such as watching television or thumping computer keys. As we shall see later, developments in technology have enabled traditional beers hitherto sold only in casks to be packaged in cans, leading to a major growth in so-called draft beer in a can. The growth in volume of NRBs, increasingly the selection in bars globally, perhaps reflects nothing more than the emerging preferences of the younger drinker, for whom the right label on the right bottle in the right hand in the right location seems to be the primary driver for beer selection. Perhaps the subconscious is also at play: the original rationale for taking one's beer straight from a personally uncorked bottle was to avoid somebody slipping you poisons!

It would be impossible in a book of this size to fully explain the evolution of the brewing industry in each of the very many countries where beer is produced. Indeed, I could devote page after page to the many political forces that have come to bear on a commodity that will always attract all shades of public opinion. A classic example is the pressure that led to Prohibition in the United States between 1919 and 1933 (see box 1.1).

Certainly, the current status quo in world brewing is in favor of huge brewing concerns: 55% of the 1.55 billion hectoliters of beer brewed worldwide in 2004 came from just ten companies (table 1.1). It is striking, too, that there are major breweries located in countries that do not have a high indigenous beer-drinking population. In France, for example, personal beer consumption is 33.4 liters per head—less than a third of the amount drunk in Germany—yet the Kronenbourg breweries (acquired several years ago from Danone by Scottish & Newcastle, a company that at the time of writing has just been acquired by a consortium of Carlsberg and Heineken) are part of a group with global sales vastly higher than Germany's biggest producer, Radeberger, which has worldwide sales of less than 15 million hectoliters. Brazil's beer consumption per capita is also far lower than that in Germany, yet it has one huge brewer called InBev, the biggest in the world and headquartered in Leuven, Belgium. It arose by a Pacman-like succession of acquisitions by Interbrew (flagship brand Stella Artois). Thus Interbrew acquired, inter alia, Labatt in Belgium, half of the original Bass company in the United Kingdom (the rest went to Coors), Whitbread of England, and Beck's in Germany. Then Interbrew merged with AmBev of Brazil (which itself had only recently been formed by the merger of Brahma and Antarctica).

The brewing industry in Germany then is characterized by many relatively small brewing companies mostly producing individual beers for predominantly local consumption. The biggest Brewer in Germany, Radeberger, produced some 14 million hectoliters for domestic consumption in 2005, and that was a third more than the next biggest competitor. There are not many truly international German brewing brands, as indeed is the case for many other countries. It is largely the brands of some of the big ten Brewers that stand alongside the great colas on the international stage, brands such as Budweiser, Heineken, and Carlsberg. Guinness is another gigantic world brand, produced by a company (Diageo) with an output of beer sufficient to make it number 15 in size worldwide.

BOX 1.1 PROHIBITION

The temperance movement began in the United States in the early nineteenth century, with 13 states becoming "dry" between 1846 and 1855, Maine leading the way. Ironically, 1846 also marked the birth of Carry Nation (1846–1911), a doyenne among Prohibitionists, whose prayers and lectures in Kansas developed into more physical acts of objection to drink when she and her followers started to smash beer containers with hatchets hidden beneath their skirts. The Anti-Saloon League was formed in Washington, D.C., in 1895, and for the first time the Prohibitionists had focus and organization. Widespread calls for prohibition were largely precipitated by claims that extensive drunkenness severely hampered productivity during the First World War. Woodrow Wilson's Food Control Bill of 1917 was aimed at saving grain for the war effort, diverting it to solid food use, and to many appeared an attempt to kill off beer. On January 26, 1920, the Eighteenth Amendment to the American Constitution was enacted. This forbade the "manufacture, sale, and transportation of intoxicating liquors" and was approved by all but two states. The Volstead Act of the same year was the basis on which the federal government enforced the block on all intoxicating liquor, defined as a drink containing in excess of 0.5% alcohol. Beer stocks were destroyed, and 478 breweries were rendered unable to go about their primary business. One of the biggest names, Lemp in St. Louis, closed their doors forever. Others developed alternative products that their technology might be turned to, such as ice cream, nonalcoholic malt-based beverages (including "near beer"), yeast, and syrups.

Carry A. Nation (photo courtesy of the Kansas State Historical Society)

(continued)

BOX 1.1 CONTINUED

Of course, though many were ardent in their antialcohol beliefs, there were those who enjoyed a drink. Unsurprisingly, the introduction of official prohibition prompted the growth in illegal home-brewing (of some dubious concoctions) and of the boot-legging/speakeasy culture colorfully portrayed in the movies. Before Prohibition, there were 15,000 saloons in New York. One year after the Volstead Act, there were more than twice as many speakeasies! Gangsters grew rich at a time when the federal authori-ties convicted 300,000 people of contravening the law. Drink-related crime surged; for example, there was an almost 500% increase in drunk-driving offenses in Chicago. People resented being prevented from partaking of something they enjoyed.

By 1933, opinion in the United States had changed (a slogan of Franklin Delano Roosevelt's Democrats was "A New Deal and a pot of beer for everyone"), and the pas-sage of the Cullen Bill allowed states not having local prohibition laws to sell beer con-taining 3.2% alcohol by weight. On April 7, 1933, the Twenty-First Amendment to the Constitution repealed the Eighteenth Amendment. Whether to enforce Prohibition or not became a state issue—but it took Mississippi until 1966 to emerge from being the last dry state. For a company to return to brewing after such a hiatus (thirteen years for most states) is no trivial issue. In particular, there had been seepage of trained and skilled brewers and operators and an inheritance of unreliable equipment, leading to equally "dodgy" products in many instances. It was the strong and the resourceful that survived, and inevitably this meant strength in size.

The United States of America is not the only country to have embraced Prohibition—you can go back as far as Egypt 4,000 years ago to find the first attempts to control the sale of beer, it being felt even then that drinking interfered with productivity. Strong temperance movements grew up in Great Britain, largely in response to the perceived excesses of drink in the burgeoning industrial cities. People were urged to sign a "pledge" not to drink, but for many the soul was weaker than some of the ale! As recently as the 1950s, in Canada one was obliged to purchase an annual permit to acquire alcoholic drinks. Prohibition was total in Finland for exactly the same period as in the United States.

A particularly vigorous temperance campaign was waged in New Zealand in the nine-teenth and early twentieth centuries, which was perhaps ironic insofar as the first alco-holic drink brewed in New Zealand was by the Englishman who "discovered" that land in 1769, Captain James Cook. A referendum after the First World War, which ended 51:49 in favor of "continuance" of the liquor trade (thanks largely to the vote of the military), enabled the beer business to continue.

In Australia during the nineteenth century they had the peculiar concept of the *six o'clock swill*. The bars opened at five and closed at six. So when the workers clocked off at five, they dived into the pubs, where filled glasses were already set up on the coun-ter. These were devoured and rapidly refilled by the bartender hosing in the beer from a tap somewhat like the hose on a gasoline pump. In this way the drinkers were able to get their fill (and more) in short order. And so a desire to limit consumption actually had the impact of encouraging people to binge-drink. There was no realization that if the restrictions had not been placed, there would not have been an encouragement to abuse.

Perhaps the most curious of the "anti-drink" movements was that in Germany in 1600. The Order of Temperance said that adherents should drink no more than seven glasses of liquor at one time and that there should be no more than two such sessions each day!

TABLE 1.1. THE WORLD'S BIGGEST BREWERS

Company	Country*	Worldwide sales in 2006 (million hectoliters, approx.)
InBev	Belgium	222
SABMiller	U.K.	216
Anheuser-Busch	U.S.A.	183
Heineken	Netherlands	132
Molson-Coors	U.S.A.	50
Modelo**	Mexico	49
Carlsberg	Denmark	49
Tsingtao***	China	46
Baltika+	Russia	45
FEMSA++	Mexico	38
Yanjing	China	38
Scottish & Newcastle	U.K.	30
Asahi	Japan	24
Kirin	Japan	24
Diageo	Ireland	19
Efes	Turkey	19
Schincariol	Brazil	18
Chongqing	China	17
Polar	Venezuela	17
Gold Star	China	17

*   Location of company headquarters
**   50% Anheuser-Busch
*** 27% Anheuser-Busch
+   50% Carlsberg and 50% Scottish & Newcastle
++ 8% SABMiller

There has been tremendous rationalization in the brewing industry in all countries, with bigger and bigger volumes being concentrated in fewer, larger companies. It almost seems as if the top ten of Brewers is just as eagerly contested as the pop charts or a football (aka soccer) league table. In the previous edition of this book I wrote that "it is only a matter of time before (anti-monopoly laws permitting) there will be mergers between those within the top 10 making, perhaps, four mega Brewers." Comparison, then, of the top ten in 2000 and 2003–2004 illustrates powerfully how this came to pass. Interbrew merged with AmBev to make InBev; South African Breweries acquired Miller to make SABMiller; Molson merged with Coors, making for a new name in the top 10 (neither company having been there before). Anheuser-Busch was pushed into third place on a global volume basis. What next? It had been rumored that Anheuser-Busch would partner with Diageo, brewers inter alia of Guinness, Harp, and Red Stripe, but also many other alcoholic beverages, from Bell's and Bushmills to Blossom Hill and Baileys and from Dom Perignon to Captain Morgan. But as I write it seems certain that Anheuser-Busch will be bought by InBev, there being little overlap in their markets. The reality is that InBev, in the guise of Labatt, has long since brewed Budweiser under license from Anheuser-Busch and it constitutes the biggest beer brand in Canada. Meanwhile, Anheuser-Busch imports a plethora of InBev brands into the United States, including Becks, Stella Artois, and Bass. (Even more remarkable is Anheuser-Busch's import agreement with another company, Budějovický Budvar, to import to the United States a beer called…Budweiser. The two companies have quarreled around the world about the right to use that name for their beer.)

As shown in table 1.2, beer production and consumption statistics differ enormously between countries. The growth in the Chinese beer industry represents a remarkable march (table 1.3), and now it is comfortably the biggest market worldwide, followed by the United States.

The Czech Republic lays claim to the highest per capita consumption of beer, 42.2 liters more passing down the throats of each Czech than those of their nearest challengers, the Germans. By contrast, the Chinese drink only 22.3 liters per head. This reflects the purchasing power of many in that country.

In some countries the brewing industry has been depressed, with a substantial decline in production volumes. Traditional beer-drinking nations such as the United Kingdom, Germany, and Denmark, even the Czech Republic, are all in decline. In part this reflects a tightening of the belt of the consumer and perhaps a change in drinking habits, but it is also recognized that drinking of even moderate amounts of alcohol is unacceptable if one is also to participate in other activities, notably driving. Countering this, however, is the increasing realization that moderate consumption of alcohol is beneficial to health.

In truth, the drinker is often never quite sure who owns the brand in her glass. In Finland, Carlsberg owns the other major brewer, Sinebrychoff, and they have a major presence in Sweden. In the Czech Republic the famed Pilsner Urquell, the original brand of the genre, is owned by SABMiller, while the second largest

TABLE 1.2. WORLDWIDE BREWING AND BEER STATISTICS (2004)

Country	Population (mill)	Production (m hl)	Imports (m hl)	Exports (m hl)	Consumption (l per head)	draft (%)	Av. strength % ABV
Argentina	37.9	13.2	0.019	0.16	35.6	–	–
Australia	20.3	17.6	0.171	0.13	86.8	19	4.2
Austria	8.21	8.78	0.56	0.58	108.6	30	4.9
Belgium*	10.9	17.8	0.99	8.7	93.3	36	–
Brazil	179.1	82.6	0.04	0.28	46.6	4	–
Bulgaria	7.8	4.8	0.023	0.013	61.0	–	4.8
Canada	31.9	23.1	2.16	3.65	67.7	10	–
Chile	16.0	4.19	0.08	0.02	27.0	–	–
China	1,306.9	291.0	1.21	1.41	22.3	2	–
Colombia	45.33	12.83	0.049	0.034	29.0	–	–
Croatia	4.4	3.6	0.302	0.292	81.2	–	–
Cuba	11.3	2.49	0.043	0.005	22.4	–	–
Czech Republic	10.2	18.8	0.202	2.7	158.0	50	4.5
Denmark	5.4	8.5	0.404	4.21	89.6	13	4.6
Finland	5.3	4.6	0.32	0.23	84.0	18	4.5
France	60.6	16.8	5.5	2.0	33.4	24	5.0

(continued)

TABLE 1.2. CONTINUED

Country	Population (mill)	Production (m hl)	Imports (m hl)	Exports (m hl)	Consumption (l per head)	draft (%)	Av. strength % ABV
Germany	82.5	106.2	3.1	13.8	115.8	20	4.8
Greece	11.1	4.2	0.53	0.15	41.2	4	4.9
Hungary	10.1	6.8	0.97	0.05	77.2	15	4.8
Ireland	4.0	5.2	–	–	108.0	72	4.2
Italy	58.1	13.2	4.87	0.85	29.6	14	5.0
Japan	127.7	66.0	0.33	0.20	51.3	18	–
Korea (Rep.)	47.9	20.2	0.213	0.57	41.4	–	–
Mexico	106.2	68.5	0.944	14.5	51.5	–	–
New Zealand	4.1	3.06	0.219	0.13	77.0	30	4.0
Netherlands	16.3	23.8	1.83	13.0	77.9	29	5.0
Nigeria	127.8	9.4	0.015	0.011	7.4	–	–
Norway	4.6	2.2	0.21	0.03	55.0	23	4.3
Peru	27.5	6.1	0.009	0.037	22.6	–	–
Philippines	81.5	13.0	0.004	0.126	15.8	–	–
Poland	38.2	31.9	0.17	0.60	82.0	11	–
Portugal	10.1	7.4	0.119	1.26	61.7	20	5.1

Romania	21.7	0.058	0.024	66.7	15	4.8
Russia	143.2	2.175	1.9	59.7	6	–
Slovak Republic	5.4	0.4	0.18	82.4	43	4.6
Slovenia	2.0	0.135	0.53	80.5	31	–
South Africa	46.4	0.7	0.2	54.4	1	5.1
Spain	43.1	3.5	0.81	77.6	30	5.2
Sweden	9.0	0.847	0.119	51.4	13	4.0
Switzerland	7.3	0.70	0.022	57.3	31	–
Ukraine	47.3	0.025	0.2	40.6	–	–
U.K.	59.8	7.0	4.36	99.0	56	4.2
U.S.A.	293.7	30.0	4.4	81.6	9	4.6
Venezuela	26.2	0.003	0.11	84.2	–	–

* Includes Luxembourg, because of inaccuracies introduced by cross-border trading

Source: Statistical Handbook, Brewers and Licensed Retailers Association, London, 2006

TABLE 1.3. GROWTH OR DECLINE IN BEER VOLUME (MILLION HL) SINCE 1970

Country	1970	1980	1990	1995	1999	2004
Denmark	7.1	8.2	9.0	10.1	8.0	8.5
France	20.3	21.6	21.4	20.6	19.9	16.8
Germany	103.7	115.9	120.2	117.4	112.8	106.2
Ireland	5.0	6.1	6.4	7.4	8.6	5.2
Netherlands	8.7	15.7	20.0	23.1	24.5	23.8
U.K.	55.1	64.8	61.8	56.8	57.9	57.4
South Africa	2.5	8.3	22.6	24.5	25.6	24.4
China	1.2	6.0	69.2	154.6	207.4	291.0
Japan	30.0	45.5	66.0	67.3	72.2	66.0
Korea (Republic of)	0.9	5.8	13.0	17.6	14.9	20.2
Australia	15.5	19.5	19.4	17.9	17.3	17.6
Canada	15.8	21.6	22.6	22.8	23.0	23.1
United States	158.0	227.8	238.9	233.7	236.5	233.2
Brazil	10.3	29.5	58.0	84.0	80.4	82.6
Mexico	14.4	27.3	39.7	44.5	57.3	68.5
World	648.1	938.6	1,166.0	1,249.8	1,367.3	1,554.6

Source: *Statistical Handbook,* Brewers and Licensed Retailers Association, London, 2006

brewer, Staropramen, is owned by InBev. Modelo in Mexico is 50% Anheuser-Busch, while FEMSA is 8% SABMiller, which owns Peroni in Italy. In New Zealand, while 33% of DB Breweries is Heineken money, Kirin has a 45% stake in Lion Nathan. In turn, Lion Nathan owns a diversity of breweries in Australia, notably Perth-based Swan, Brisbane's Castlemaine, and Tooheys from Sydney.

At the other extreme, there has been a gratifying trend in the establishment of newer, smaller breweries, called either microbreweries or pub breweries depending on their size. The Institute for Brewing Studies defines a microbrewery as one that produces less than 15,000 barrels of beer each year. A brewpub is classified as an establishment that sells the majority of its beer on-site, whereas a contract brewing company is a business that hires another company to produce its beer. A prominent example would be the Boston Beer Company, although more recently they have been focusing on building their own breweries, for example in Cincinnati. A regional brewery has a capacity of between 15,000 and 2 million barrels.

In the United States in the 1960s there were fewer than 50 breweries. Now (as I write) there are 1,449—and they're still coming, many with a capacity of just a few barrels. President Jimmy Carter's initiative to allow home brewing in the United States was one significant factor presaging the start of the microbrew surge, as was the entrepreneurial and devoted attention of pioneers such as Fritz Maytag, in his case with the Anchor Brewing Company in San Francisco (see box 1.2). The sector generates a healthy consumer interest

BOX 1.2 ANCHOR

Let's go back to the year 1965. The *San Francisco Examiner* declared that Haight-Ashbury was the New Bohemia. The Grateful Dead first saw life. And a 28-year-old man sitting in the Old Spaghetti Factory in North Beach fell in love with a beer by the name of Anchor Steam. The man was told that if he liked the beer so much he had better get down to the brewery, because it was about to close. The man was Fritz Maytag. He bought the company. He turned it around. And he is now solidly regarded as the messiah of the brewing revolution in the United States.

Frederick Louis Maytag III was born in Iowa, in Newton, the county seat of Jasper County. His great-grandfather F. L. Maytag I founded the appliance company, which these days is publicly owned. Grandfather Elmer Henry Maytag established a herd of Holstein cattle in order to furnish the family and workforce with high-quality milk, and it was his sons F. L. Maytag II and Robert who in 1941 established the dairy that converted this milk into the famous Blue Cheese, a product that nowadays includes Elton John and Oprah Winfrey among its aficionados.

The Maytag family clearly recognized the merits of the land grant university system, and it was Iowa State that came up with the process that Maytag Dairy Farms employed to make blue cheese from cow's milk rather than the traditional sheep's milk.

Young Fritz was educated in the local public school through ninth grade. To quote Fritz from an interview in the *San Francisco Chronicle*, "My parents were extremely desirous of us having a normal life, not feeling that we were anything different than anyone else as little children, and having a sense of humility that we had a famous name and half the town worked for our father. We were taught to realize that we were not anything special at all."

He graduated from the Deerfield Academy in Massachusetts in 1955, an institution that counts Matthew Fox, King Abdullah II of Jordan, and diverse members of the Bush family as its alumni.

As regards what happened next, Fritz is variously quoted as saying "I had seen the east and I didn't like it" and "My sister had chosen Stanford…they wore Levis to class and there were girls everywhere…" In any event that's where he went, as a liberal arts student.

Fritz studied a diversity of subjects, and developed a passion for Chinese and Japanese literature. Soon, though, he found his way to San Francisco, and the rest, as they say, is history.

It was Ralph Waldo Emerson who said, "God made yeast, as well as dough, and loves fermentation just as dearly as he loves vegetation." There is only one man I can think of who truly embodies this credo, Fritz Maytag.

He has the brewery on Mariposa Street in Potrero Hill, the 13th biggest brewing company in the United States. Once he worked with a noted historian to re-create a

(continued)

BOX 1.2 CONTINUED

beer according to the methods employed in Ancient Sumeria of 6,000 years ago. It was launched at a banquet in San Francisco, with everyone sipping through straws in the old way. Fritz will tell you it wasn't very good—but that certainly isn't the case for his real beers, which include such magnificent brews as Anchor Steam, Liberty Ale, Old Foghorn, and 30 years' worth of eagerly anticipated Christmas ales.

He is chairman of the board of the cheese company back in Iowa.

He founded the Anchor Distilling Company in 1993, whose products include the single-malt rye whiskey Old Potrero and Junipero Gin.

He owns the York Creek vineyards, having developed an interest in wine during his humanitarian work in Chile.

Fritz even makes apple brandy from his own fruit and presses his own olive oil for private consumption.

And to all of these he brings the same philosophy of quality. Nothing but the best. And a spirit of helping others. And so, in brewing, he has always been at pains to help other people, recognizing that it takes more than just one brewer to satisfy the demand for a beverage of excellence. When he acquired Anchor, the number of non-transnational brewing companies in the United States could be counted on the fingers of one, possibly two hands. Now there are more than 1,400 breweries in this country. If you took a straw poll of the names responsible for this astonishing surge in interest, you would find most people listing Fritz Maytag as the key player.

Somehow, Fritz Maytag finds time to indulge hobbies, which include astronomy, desert plant life, and driving trotting horses. And I know that somewhere he has a rather natty little red sports car of British heritage.

in beer and in the art of brewing. Whether on street corners, dispensing full-flavored beers of diverse character to accompany economically priced meals, or in baseball stadiums, adding to the sublime pleasure of the ball game, these tiny breweries greatly enrich the beer-drinking culture. Each year a succession of young hopefuls steps into my office expressing their overwhelming desire to open a brewery. A laudable sentiment indeed. My reply to them is invariably that, apart from my class, what they need is a compassionate bank manager and a chef, for let nobody kid themselves: it is the food that pays the bills in a pub brewery, more than the beer. It is only when a company gets to the size of the likes of Anchor (more than 100,000 barrels per annum) and most exceptionally Sierra Nevada (see box 1.3) that beer is unequivocally the driving force. Smaller-scale operations are starting and finishing all the time. The

BOX 1.3 SIERRA NEVADA

Ken Grossman commenced his brewing (as so many folk do) in a bucket at home, in his case in Southern California. In 1972 he moved north to study science at Butte College and California State University, Chico. Four years later he opened a small shop in Chico selling home brewing supplies (he had previously run a bicycle repair business), while daydreaming about opening his own commercial brewery, inspired by the pioneering efforts of Jack McAuliffe and his New Albion brewery in Sonoma. In 1980, the Sierra Nevada brewery was opened in a small warehouse in the city by Grossman and partner Paul Camussi, with converted dairy equipment and a packaging line converted from soft drink use. The brewery was named for Ken's favorite hiking grounds. Ken tells of how he enrolled in every class at Butte College that featured time in the mechanical shop. Imagine this scenario: a man studying by day and going home to make his own brewing equipment, including drilling hundreds and hundreds of holes in the base of a tank to make his own lauter tun. And then, having made the beer, wandering the streets of Chico asking the bar owners if they would stock it.

Soon the prizes started accumulating for a series of distinctively hoppy beers, including the flagship Pale Ale in its distinctive bottle. Now, a little more than 20 years

Early days for Sierra Nevada—Ken Grossman shifting cases of beer (photo courtesy of Sierra Nevada Brewing Company)

(continued)

BOX 1.3 CONTINUED

later, the company is producing more than 700,000 barrels of beer each year, shipping it to every state, and operating one of the most impressive and delightful breweries to be found anywhere in the world, the seventh biggest brewing company in the United States. It is as esthetically pleasing as it is technologically state-of-the-art. There are murals on the brewhouse wall, individual handcrafted tiles on the corridors that depict scenes from barley to beer by way of all stages in the process. Everywhere are gleaming copper and spotlessness, speaking of Grossman's passion and unstinting commitment to quality.

A tiny fraction of the hops that Ken Grossman uses are grown on site. The spent grains go to the local college farm to be fed to cattle whose destiny is steaks in the sumptuous restaurant alongside the brewery. Through its anaerobic effluent digestion plant, fuel cell, and solar power systems, that brewery is remarkably proficient from an energy perspective, gaining prestigious awards from Governor Schwarzenegger. There are the Big Room, a function facility catering for weddings and meetings, and a musical program that is Chico's answer to the Austin City Limits. Sierra Nevada has sponsored a professional bicycle racing team since 2001.

And not a single television advertisement: The beer speaks for itself and for the modest man who is now the sole owner of the company.

Sierra Nevada brewery 2008 from the air (Brian Peterson Photography, 2008)

Ken Grossman with Governor Arnold
Schwarzenegger (photo courtesy of Sierra
Nevada Brewing Company)

failure rate for brewpubs and microbreweries in the United States averages out
at around 1 in 3. But the scenario is far from gloomy overall. In 2007, the craft
brewing sector was the top performer in terms of growth in the U.S. industry,
with a 12% volume increase over the previous year. Even then, it amounts to
only 3.8% of production and 5.9% of retail sales.

Table 1.2 reveals several more intriguing statistics. For example, the average strength of beer ranges from a high of 5.2% alcohol by volume (ABV) in
Spain to a low of 4.0% in New Zealand and Sweden. This disguises, of course,
a myriad of beer types of diverse strengths, which in the United Kingdom, for
instance, ranges from alcohol-free beers (by law containing less than 0.05%
ABV) to the so-called super lagers of 9% alcohol or above. Nonetheless,
national preference is reflected in the strengths indicated in table 1.2. People
often confuse strength with flavor. Some of the British ales are relatively low

in alcohol content, despite their fuller flavor. Those used to drinking them can get a curious surprise after a few beers in, say, the United States, where many of the mainstream beers tend to be substantially stronger in terms of alcohol if not flavor intensity.

The great beer-exporting countries of the world, with the exception of Germany, feature major brewing companies. The Netherlands, home of Heineken, exports more beer than any other country, some 54% of its production. Denmark, where Carlsberg is based, exports 49% of its beer.

The export of beer first took off with British imperialism in the nineteenth century, and with the shipping of vast quantities of so-called India Pale Ale (I.P.A.), a product still available from several Brewers in the "home market" today. This beer was strong in order to suit drinkability in hotter climes, and it was well hopped, as hops have preservative qualities. The advent of pasteurization, and the attendant destruction of potential microbial contaminants, enhanced the market for such exports, as it meant that shelf lives could be lengthened still further.

Beers are still exported from country to country, a principal driving force being the opportunity to make marketing claims concerning the provenance of a product. However, most major Brewers realize how illogical it is to transport vast volumes of liquid across oceans—after all, by far the major component of beer is water! They have either established their own breweries to supply specific market regions, or have entered into franchise agreements with Brewers in target countries, who brew their beer for them, generally under extremely tight control. For example, beers from major American Brewers are brewed locally in the British Isles, with each of these companies insisting on adherence to brewing recipes, yeast strain, and the various other features that make their brands distinctive. Companies operating franchise agreements may insist on key technical personnel being stationed in the host brewery in order to maintain responsibility for a brand. A good example would be the presence of a brewer from Kirin at Anheuser-Busch's Los Angeles plant.

There are, of course, circumstances when a franchise brewing approach is impractical and when it is also not possible to ship finished product. For instance, in 1944 HMS *Menestheus* was converted from a minelayer to a floating club and brewery. Seawater was pumped on board and distilled to produce the brewing liquor. Malt extracts and hop concentrates represented transportable ingredients for a nine-day production cycle. The production rate was 1,800 gallons per day for the pleasure of faraway troops.

All Brewers are well aware of the fact that they are not only in competition with one another in the marketplace, but also with producers of other drinks, both alcoholic and nonalcoholic. The esteemed drinks analysis company

Canadean calls it "share of throat." Yet if we look at data for per capita drinks consumption in the United States in terms of numbers of servings, beer is second only to soft drinks, with coffee and milk some way behind. If you consider that the legal drinking age in this country is 21, it is clear that beer commands a significant position in the rankings of drinks purchased. In 2004 beer accounted for 53% of the total alcoholic beverage market in the United States, with spirits at 32% and (despite the hype) wine at 15%.

For wines and spirits, just as for beers, there are distinct national differences in consumption (table 1.4). In most countries more beer than wine is consumed (although we should remember that wines generally contain two to three times more alcohol than does beer, volume for volume). However, the French drink considerably more wine than beer, whereas in Portugal there is almost equivalence between the two beverages.

TABLE 1.4. DRINKS CONSUMPTION (PER CAPITA, 2004)

Country	Beer	Wine	Spirits
Australia	86.8 (−8.6)*	21.7 (+10.2)	1.6 (+14.3)
Belgium	93.3 (−7.2)	24.0 (+10.6)	1.2 (0)
Brazil	46.6 (−0.6)	2.2 (+10.0)	−
Canada	67.7 (−0.6)	11.0 (+19.6)	2.3 (+21.1)
China	22.3 (+36)	1.0 (+25)	2.8 (0)
Czech Republic	158.0 (−4.2)	16.5 (+4.4)	3.0 (−10)
Denmark	89.6 (−15.0)	30.0 (+0.7)	1.5 (+36.3)
Finland	84.0 (+4.8)	9.7 (+14.1)	2.5 (+31.6)
France	33.46 (−13.7)	48.5 (−15.2)	2.4 (0)
Germany	115.8 (−9.2)	20.1 (−12.2)	2.0 (0)
Republic of Ireland	108.0 (−14.3)	15.3 (+66.3)	2.6 (+23.8)
Italy	29.6 (+9.2)	49.8 (−3.3)	0.3 (−40)
Japan	51.3 (−9.0)	2.0 (−15)	2.4 (−4.0)
Netherlands	77.9 (−7.7)	22.1 (+18.8)	1.5 (−11.8)
New Zealand	77.0 (−4.6)	20.6 (+13.8)	1.7 (+13)
Norway	55.0 (+6.4)	15.9 (+59)	0.8 (−11)
Portugal	61.7 (−4.9)	47.6 (−16.5)	1.0 (0)
Russia	59.7 (+95.7)	8.4 (+52.7)	6.0 (0)
Slovak Republic	82.4 (−3.7)	13.4 (−11.8)	3.8 (−7.3)
South Africa	54.4 (−8.1)	6.4 (−20)	0.8 (−11)
Spain	77.6 (+12.8)	31.9 (−10.4)	1.9 (+5.6)
Sweden	51.4 (−13.3)	14.7 (−7.0)	1.3 (+18)
U.K.	99.0 (−1.4)	21.9 (+32.7)	1.9 (+18.8)
U.S.A.	81.6 (−3.3)	8.2 (+10.8)	2.0 (0)

* Values in parentheses indicate percent growth or decline since 1999. Values are given as liters or, for spirits, as liters of pure alcohol.

Source: *Statistical Handbook*, Brewers and Licensed Retailers Association, London, 2006

One significant factor influencing the respective amount of beer and wine drunk in different countries is the relative excise tax (duty) levied on them (table 1.5). In seven member states of the European Community (EC), including Italy, wine attracts no duty whatsoever. The tax levy on wine in France is very low, whereas duty rates on wine (but also on other types of alcoholic beverage) are very high in Sweden, Finland, and Ireland.

There are huge differences in the excise rates for beer across the EC. This issue has been brought to the fore in the United Kingdom, in view of the fact that France is nowadays just a 30-minute train ride away through the Channel Tunnel. As beer is so much cheaper in France because it attracts less than one-seventh of the excise duty levied in the United Kingdom, a growing number of people make trips across the English Channel to buy stocks. Well over a million pints of beer each day are coming across the Channel into England and, thence, to the rest of the United Kingdom There are no limitations on the amount of beer you can bring back to the United Kingdom, provided it is for personal consumption, but the retail of such purchases is forbidden. Yet probably half of this imported beer is intended for illegal disposal. From the

TABLE 1.5. RATES OF EXCISE DUTY AND VALUE ADDED TAX IN THE EUROPEAN COMMUNITY

Country	Beer (cents per pint at 5% ABV)	Wine (cents per 75cl bottle at 11% abv)	Spirits ($ per 70cl bottle at 40% abv)	VAT %
Austria	18.8	0	3.86	20.0
Belgium	16.0	48.8	6.76	21.0
Denmark	26.4	85.2	7.76	25.0
Finland	76.2	219.4	10.92	22.0
France	10.2	3.6	5.60	19.6++
Germany	7.4	0	5.04	16.0
Greece	10.6	0	4.20+	198.0
Ireland	77.8	282.4	15.16	21.0
Italy	22.2	0	3.08	20.0
Luxembourg	7.4	0	4.02	15.0*
Netherlands	19.6	61.0	5.80	19.0
Portugal	12.6	0	2.92	21.0*
Spain	7.8	0	3.20	16.0
Sweden	71.2	243.0	20.60	25.0
U.K.	75.4	258.2	10.96	17.5

*   VAT rate for wine is 12%; + 50% less for ouzo
++ VAT rate is 5.5% for beer sold in supermarkets and shops
Source: *Statistical Handbook*, Brewers and Licensed Retailers Association, London, 2006. Original data was quoted in Pounds Sterling. An exchange rate of £1 = $2.00 has been employed and values rounded to one decimal point.

numbers of vans returning through Kent packed to the roof with beer, it would appear that either there are some fun parties to attend in Britain, or else the law is being flouted "big-time"! Hundreds of millions of dollars' worth of tax revenue is evaded through smuggling operations into the United Kingdom. It seems unlikely that the duty imbalance will change substantially, particularly as beer duty contributes a very substantial proportion of the receipts of Her Majesty's Revenue and Customs, with a huge additional take from value added tax (VAT). No other member state of the European Community collects anywhere near as much revenue from Brewers. France, ironically, is the next biggest drawer on Brewers, but levies less than 30% of the tax taken in the United Kingdom, most of that being VAT.

In the United States matters are complicated by there being three layers of government levying taxes on beer, a relic from days of Prohibition. Congress first placed an excise tax on beer in 1862. The federal rate of excise tax for large Brewers has been $18 per barrel (U.S.) since it was doubled in 1990 following strong lobbying by antialcohol advocates. A reduced rate of $7 per barrel for the first 60,000 barrels of beer annually is provided for brewers who produce no more than 2 million barrels in a calendar year.

State excise tax varies tremendously, but the current median is 18.8 cents per gallon. Sales taxes also apply in most states. The lowest rate of taxation is in Wyoming at 2 cents per gallon, while there is a whopping $1.07 per gallon in Alaska. Other high rates are in Hawaii, Alabama, North Carolina, South Carolina, Florida, and Georgia, whereas they are somewhat low in Colorado, Maryland, Missouri, Kentucky, Oregon, Pennsylvania, Wisconsin, and the District of Columbia.

Although production costs associated with the brewing industry vary enormously from company to company, I would estimate that excise tax probably accounts for approximately 27% of the cost of beer in the United States. Estimates for other outgoings would be malt (3.5% of costs), adjuncts (1.5%), hops (0.2%), packaging materials (26%), production costs (20%), and sales costs (21%). Hence, excise duty is one of the single most costly elements of a can of beer.

America's beer industry, from suppliers through brewers to distributors and retail contributes almost $190 billion annually to the United States. This includes more than 1.7 million jobs and more than $36 billion in federal, state, and local taxes.

Despite the competition that beer faces from wine, there has been a steady growth in world beer production in recent years, and this growth is projected to continue. The volume of beer brewed has increased 2.4-fold since 1970, in which time the world population has increased by 75%. As tables 1.3 and 1.4 show, there has been formidable growth in the quantity of beer brewed and consumed in a number of countries. China stands out as a nation where an increasing number of people in an increasingly favorable economic climate have acquired access to beer, but at least as remarkable in the past several years has been the surge in the brewing industry in Russia, reflecting a shift from vodka to the drink of moderation.

Returning to the United States, and before we leave this statistical survey, we might analyze the drinking habits of the individual states of the Union (table 1.6). It seems that the good folk of Nevada head up the beer stakes, with New Hampshire close behind. Utah, unsurprisingly, has the lowest per capita consumption.

With the exception of brewpubs, where the beer is brewed on the premises, it is illegal for a brewer in the United States to sell directly to the consumer. This is quite unlike the situation in some other countries, where brewers to a greater or lesser extent are able to act as sellers as well as producers, for example through their own pubs.

After the repeal of Prohibition, the center of gravity on control of beer sales was placed firmly on the states rather than on the federal government. The upshot of this is that there is a plethora of differences, some subtle but some substantial, in the laws governing beer in different locations. One of the most significant comes in Utah, where by law no beer for sale can be in excess of 3.2% alcohol by weight (4% ABV), unless it is retailed in state-owned stores. Some states will allow beers of typical average strength (e.g., 5% ABV) but prohibit the sale of the very strong beers.

In all states there is, by law, a three-tier arrangement comprising the Brewer (supplier), the distributor (wholesaler), and the retailer. The majority of distributors will deal with products from different Brewers. The system hinges on the negotiated contracts drawn up between the various elements of the chain, which will embrace not only cost structures but also quality-related issues such as hygiene standards, age of beer in storage, and so on.

A BRIEF HISTORY OF BEER

As this small volume is published in the United States, whose brewing pedigree has British and then German antecedents, I will focus on the history of brewing in those two countries, after a brief look at the very roots of beer and before addressing the relatively youthful history of the product Stateside.

ANCIENT ORIGINS

Beer probably originated approximately 8,000 years ago in Sumeria, within the Fertile Crescent (modern-day Iraq), although it has recently been claimed that

TABLE 1.6. CONSUMPTION OF BEER PER CAPITA (GALLONS)

State	2007
Alabama	20.3
Alaska	24.5
Arizona	26.1
Arkansas	19.1
California	19.6
Colorado	23.8
Connecticut	17.4
Delaware	24.1
District of Columbia	27.7
Florida	24.7
Georgia	21.1
Hawaii	25.0
Idaho	19.9
Illinois	22.9
Indiana	20.0
Iowa	22.5
Kansas	18.8
Kentucky	18.8
Louisiana	25.9
Maine	21.1
Maryland	18.8
Massachusetts	20.4
Michigan	21.2
Minnesota	21.8
Mississippi	23.7
Missouri	23.2
Montana	26.3
Nebraska	23.8
Nevada	33.2
New Hampshire	30.7
New Jersey	18.4
New Mexico	26.8
New York	17.5
North Carolina	20.1
North Dakota	25.2
Ohio	22.7
Oklahoma	19.8
Oregon	21.7
Pennsylvania	22.1
Rhode Island	21.6
South Carolina	24.4
South Dakota	23.4
Tennessee	21.1
Texas	26.8
Utah	12.4
Vermont	22.6

(continued)

TABLE 1.6. CONTINUED

State	2007
Virginia	20.6
Washington	20.2
West Virginia	20.9
Wisconsin	27.6
Wyoming	23.4
Total	21.8

Courtesy: The Beer Institute

a forerunner of beer was being brewed in Amazonia some 2,000 years before that. Beer was consumed throughout the Middle East, but as a drink it would hardly have borne much resemblance to what most of the world today regards as beer. According to Delwen Samuel, a distinguished researcher from the University of Cambridge, England:

> Beer, together with bread, was the most important item in the diet of the ancient Egyptians. Everyone, from Pharaoh to farmer, drank beer and no meal was complete without it.... [B]eer was much more than just a foodstuff. In a cashless society it was used as a unit of exchange, its value fluctuating just as currencies do today.... Furthermore, beer played a central role in religious belief and ritual practice. Offerings to the gods or funerary provisions included beer, either real or magical.

Samuel's archaeological pursuits have unveiled the remains of beer solids crusted to the inside of ancient vessels, and among these solids were found fragments of grain. She has painstakingly examined these remains using techniques such as scanning electron microscopy and has made proposals as to how beer was brewed in Egypt 3,000 years ago from malted barley and a primitive type of wheat called emmer, a process that the Egyptians adopted from the original brewmasters of Sumeria and Babylon. Indeed, the original Sumerian approach, as written down in the ancient Hymn to Ninkasi, was taken by Fritz Maytag, collaborating with historian Solomon Katz, in the production of such a beer.

The techniques applied in the brewing of beer by the Egyptians seem to have been quite refined. Quite how the first beer was developed several thousand years prior to this is unclear, but it might be anticipated that its origins were founded on serendipity and were linked to the baking of bread. Most commentators suggest that batches of barley must have gotten wet through inadequate storage (rain was more plentiful thereabouts than it is now) and, as a result, they started to germinate. Presumably it was found that drying stopped this germination, and, logically, the ancients would have discovered that this "cooking" improved the taste of the grain. Neither would it have taken them long to realize that malt is more nutritionally advantageous than raw barley: those eating malt would have been healthier than those whose diet included barley, and, for certain, they would have found their meals to be tastier.

It is supposed that the sprouted barley (forerunner to today's malt) was made into dough before bread making, and then batches of the dough spontaneously fermented through the action of yeasts living on the grain and in cracks and crevices in vessels. Soon the ancient brewers will have realized that the dough could be thinned with water and strained as a precursor to fermentation and that the process could be accelerated by the addition of a proportion of the previous "brew." A range of plants will have been used to impart flavors, among them the mandrake, which has a flavor much like leek. The use of hops came much, much later.

The work of archaeologists has suggested that in Mesopotamia and Egypt the characteristic tool of the brewer was an earthen vat. Certainly, hieroglyphics depict people stooped over such vessels in pursuit of their craft. It has been suggested that the Pharaoh Ramses had a brewery that furnished 10,000 hectoliters of beer each year free of charge to those employed in the temple. Beer was staple stuff: the Code of Hammurabi, 1,800 years before the birth of Christ, decreed that those overcharging customers for their beer were to be drowned.

It has even been claimed that modern civilization has its origins in the brewing of beer and that the urge to domesticate barley and cultivate it in a controlled manner for the production of beer was the justification for our ancient forebears settling in communities rather than pursuing a nomadic existence.

The Egyptians passed on their brewery techniques to the Greeks and Romans. However, in ancient Greece and Rome wine was the drink of the privileged classes, with beer consumed by the rest. Beer was not foremost amongst the developments bestowed by the Romans in the lands that they conquered. Pliny the Elder (23–79 A.D.), a Roman author, was almost contemptuous in his view that

> People who live in the west have their own form of intoxicating drink. There are different techniques used in the various districts of Gaul and Spain, and the drink goes by different names, but the basic principle is the same: grain is steeped in water. . . . God knows, we used to think that Egypt produced grain for bread; but men's vices are wonderfully resourceful, and a way has been found to make even water intoxicating!

It seems that it was through a more northerly route that the Celts brought westward their ability to brew. Perhaps this related to the mastery over wood of the people of Northern and Central Europe, and their ability to fashion it into brewing vessels and barrels. Whereas the Greeks and others in the South were drinking wine from pottery, the German tribes were drinking barley- or wheat-based drinks out of wood. Pliny encountered *cerevisia* in Gaul and *ceria* (*ceres*) in Spain, hence the brewing yeast named *Saccharomyces cerevisiae*.

In the first century A.D. there is mention of Britons and Hiberni (Irish) making *courni* from barley, which had probably been cultivated in England since 3000 B.C. The Old Irish name for ale was *coirm*.

For the Danes and the Anglo-Saxons, ale was a favorite beverage, for grapes could not be readily cultivated in the colder northern climes. Beer was deemed the perfect beverage of heroes, and Norse seafarers were fortified in battle by the thought that, should they perish, they would go to drink ale in Valhalla. The Scandinavian word *bjor* is related to English *beer*.

BRITAIN BASICALLY

The manner by which the ancient Britons produced their beer is not entirely unrecognizable:

> The grain was steeped in water and made to germinate, by which its spirits were excited and set at liberty; it was then dried and ground, after which it was infused in a certain quantity of water and, being fermented, it became a pleasant, warming, strengthening, and intoxicating beverage.

Much of the history of the brewing industry is tied up with the Church, to the extent that the monks in the Middle Ages were even convinced that the mortar used in the building of their churches and monasteries was better if mixed using ale rather than water. To this day, the strong Trappist beers of Belgium are brewed by monks, and bone fide travelers in England are still entitled to lay claim to ale and bread if they care to visit a cathedral church. In 1286 the monks of St. Paul's Cathedral in London brewed almost 70,000 gallons (U.K.) of ale. Monks used the symbols X, XX, and XXX as symbols to detect sound quality in beers of increasing strength.

Ale was exceedingly popular. William of Malmesbury said of the British in the early twelfth century:

> Drinking was a universal practice, in which occupation they passed entire nights as well as days. They consumed their whole substance in mean and despicable houses, unlike the Normans and French, who in noble and splendid mansions lived with frugality. They were accustomed to…drink till they were sick. These latter qualities they imparted to their conquerors.

The monasteries passed on their skills to those brewing in their own homes. By the Middle Ages, ale had become the drink of choice for breakfast, dinner, and supper. Tea and coffee, of course, hadn't arrived.

Out of domestic brewing developed the forerunner of the brewpub, with beer brewed in the back and sold out front. The two main products were those fermented from the first strainings from the mash tun (strong beer) and those derived from the weaker later runnings (small beer). Brewing was in wooden vessels, except for an open copper for boiling the wort.

Through the Assize of Bread and Ale in 1266, aleconners were appointed in boroughs and cities to test the quality of the ale and the accuracy of the measures being used. Licenses to brew were needed as early as 1305. The aleconner wore leather breeches and would arrive at the brewery uninvited, pour a glass of ale onto a wooden bench, and sit for 30 minutes. He would smoke and drink, but otherwise remain static. Woe betide the brewer if he stuck to the seat because of sugar left unfermented in the ale.

The brewer had to put out a pole, with attached bush or ivy plant, to register that the beer was ready. Later this became a metal hoop, and various things were displayed in it to differentiate breweries. At first these were the actual objects, for example crossed keys, and the drinking house in that case would be recognized as the Cross Keys. Later, real objects were replaced by paintings, allowing for more than one pub to be called The King's Head!

By the early fourteenth century there was one "brewhouse" for every 12 people in England. The beer was brewed by females (*brewsters* or *alewives*). A *hukster* was a woman who retailed ale purchased from a manufacturing ("common") brewer, while women who sold wine were called *hostesses*.

It was frequent practice to spice ale, by adding pepper or other stimulants, to give the product an additional bite, but for long enough these flavorants did not include hops. Hops for brewing may have been first brought to the United Kingdom to satisfy retainers of Philippa of Hainault, the wife of Edward III, or by Germans to gratify the German mercenaries supplementing the British army. The cultivation of hops in the Hallertau region of Germany is first recorded in 736, and St. Hildegard, writing in 1079, is perhaps the first to have mentioned the preservative properties of this plant. It has been claimed that hops were cultivated in Kent by 1463, while some insist that the first reference to hops in England is in a document from 822 by Abbot Adalhard of Corvey. The exact provenance of the arrival of hops is, then, uncertain, but there are several versions of one particular rhyme:

Hops and turkies, carps and beer
Came to England all in a year

and

Turkeys, carps, hops, piccarel, and beer
Came into England all in one year

or

Hops, reformation, bays, and beer
Came into England all in one year.

Prior to the arrival of hops, ale had sometimes been preserved with ground ivy. There was a very clear distinction in the fifteenth century between brewers of ale and beer. In those times, the term "ale" strictly described an unhopped product, whereas beer was hopped. We British have always been a tad xenophobic, and thus for 150 years beer brewing was deemed the domain of foreigners, and ale brewers never passed up a chance to persecute them and rubbish their products. During the reign of Henry VIII, one owner of an ale brewery successfully brought an action against his brewer for putting in "a certain weed called a hop." It was decreed that neither hops nor brimstone were to be put into ale. We can be thankful that hops gained ascendancy, for they seem infinitely preferable to materials such as wormwood, gentian, chicory, or strychnine that were sometimes employed.

By 1576 beer was so prized over ale that Henri Denham, writing in *A Perfite Platforme of a Hoppe Garden,* said:

> Whereas you cannot make above 8 or 9 gallons of indifferent ale out of one bushell of mault, you may draw 18 or 20 gallons of very good Beere, neither is the Hoppe more profitable to enlarge the quantity of your drinke than necessary to prolong the continuance thereof. For if your ale may endure a fortnight, your Beere through the benefit of the Hoppe shall continue a moneth, and what grace it yieldeth to the teaste, all men may judge that have sense in their mouths—here in our country ale giveth place unto Beere, and most part of our countrymen do abhore and abandon ale as a lothsome drink.

Gerard, writing in 1596, was of the opinion that

> The manifold virtues in hops do manifestly argue the wholesomeness of beere above ale, for the hops rather make it a physical drink, to keep the body in health, than an ordinary drink for the quenching of our thirste.

An early attempt, then, to position beer on a health-positive platform.

By the middle of the nineteenth century, almost 22,000 tons of hops were grown in England, and it was referred to as the "English narcotic" as a result of its surpassing tobacco in amounts consumed.

Henry VI had appointed surveyors and correctors of beer-brewers, whose principles of operation were laudable:

> Both the malt and hops whereof beer is made must be perfect, sound and sweet, the malt of good sound corn—to wit, of pure barley and wheat—not too dry, nor rotten, nor full of worms, called wifles, and the hops neither rotten nor old. The beer may not leave the brewery for eight days after brewing, when officials should test it to see that it is sufficiently boiled, contained enough hops and is not sweet.

The reader should realize, then, that brewing has a long tradition of high standards. The longevity of the process and the fact that the unit stages of brewing have remained essentially unchanged for hundreds (if not thousands) of years will be appreciated from this description of a brewery from 1486:

> One London brewery included a copper brewing-kettle, a mashvat with a loose bottom and a tap through of lead, a vat and two kettles for wort, two leaded systems for "licuor," twenty little tubs for yeast, a fan for cooling the wort, a malt mill, twenty four kilderkins and a beer dray with two pairs of wheels.

A government Act of 1604 required parish constables to inspect alehouses to ensure that they were operated properly. (William Shakespeare's father had been an ale taster in Stratford-upon-Avon prior to this time.) It was emphasized that

> Whereas the ancient true and principal use of Inns Alehouses and Victualling-houses was for the receipt relief and lodging of wayfaring people travelling from place to place, and for such supply of the wants of such people as are not able by greater quantities to make their provision of victuals, and not meant for entertainment and harbouring of lewd and idle people.

No workman was allowed to spend longer than one hour in an inn unless occupation or residence compelled him to do so. Yet by the reign of Queen Elizabeth I it was reported that in my native Lancashire the alehouses were so crowded on a Sunday that there was nobody left in the church save the curate and his clerk.

By 1688, more than 12 million barrels of beer were drunk in a year in Great Britain, for a population of 5 million. Even infants, who drank small beer, scarcely ever drank water.

It didn't take terribly long for those in authority to realize that good steady income was to be had by taxing brewers. In 1614 James I had levied four pence per quarter of malt, while the Parliamentarians, not noted drinkers and certainly in need of revenue, imposed a duty of two shillings per barrel on beer retailing in excess of six shillings. Additional duty was placed on malt from 1697 and on hops from 1711. The first laws were already in place in various regions to reduce habitual drunkenness: these included fixed hours of closing at night, Sunday closing, and a requirement that no drinker stay longer than an hour at a time.

There were three main categories of beer: ale (strong), beer (weak), and the better quality "twopenny." There were brown, pale, and amber versions of each. People usually asked for "half and half"—equal measures of ale and beer—or "two thirds" ("three threads"): ale, beer, and "tuppenny." One story has it (but later we will hear a different version) that in 1722 a London brewer called Ralph Harwood conceived of a product analogous to two thirds, in which the three beers were premixed in the brewery, thereby saving the landlord's and the customers' time. Because most of the customers were porters in the local markets it became known as *porter*. Yet within a century the growth of porter had subsided, paler products gaining the ascendancy.

British beer was becoming popular around the world: it was being delivered to ports far and wide by proud ship captains. The Trent Navigation Act of 1699 opened up transport from Burton-on-Trent through Hull to the world—enabling the likes of Allsopp and Bass to become household names far from home base. Peter the Great and Catherine the Great in Russia were said to relish the ales shipped to St Petersburg.

Yet in early eighteenth-century London, gin was developing popularity. In 1714 there were 2 million gin distilleries in England, and 21 years later 5 million. A license was needed for selling beer, but not gin.

Benjamin Franklin wrote of the drinking habits of employees in the British printing industry: a pint before breakfast, another with breakfast, a pint between breakfast and dinner, one more at dinner, a pint at 6 o'clock, and a last one at knocking-off time. Then it was time to go out and enjoy oneself, presumably down at the pub.

Toward the end of the eighteenth century, the impact of taxation and increasing imports of tea and coffee saw a change in domestic drinking habits—tea instead of ale for breakfast.

In the late 1700s, there was a decline in beer brewed at home, reflecting the growth in towns and industry and increase in proportions of people working in factories. The development of roads and railways allowed big brewers to distribute their products. By 1815, Barclay Perkins was brewing over 300,000 barrels of beer a year in London, using the latest steam engines, invented by Richard Trevithick and James Watt, which facilitated the Industrial Revolution.

There were separate rates of excise for strong beer and small beer. Disputes aₛ to whether a beer was one or the other were settled by dipping a finger, before John Richardson constructed the first brewer's saccharometer in 1784. Twenty-two years earlier, another English brewer, Michael Combrune, had been the first to apply the thermometer in the control of a brewery's operations. Before that time it had been standard practice to poke one's thumb into the boiled wort to ensure that it wasn't too hot to accommodate the yeast. Alternatively if one boiled the water, then, when the steam had cleared, the water was at the correct temperature for use when you could see your face reflected in it.

In the late eighteenth century, the tied-house system was started in Britain, in which major production brewers sold their products through their own wholly owned pubs. By 1810, there were 48,000 alehouses for some 8 million people. Captains of the booming Industrial Revolution were concerned about wages being "wasted" on excess drinking. As a result pubs were limited to strict opening hours, which have been relaxed only very recently. The first tee-total pledge was signed in Preston in 1832.

In the nineteenth century, there was an impressive selection of beers available to the English consumer. In 1843, Burton Ale had original extracts between 19.25 degrees Plato (P) and 30°P, while common ale was 18.25°P and porter 12.5°P. (As a rule of thumb, a beer with an original extract of 10°P will give beer of 4% alcohol by volume. So one of 30°P, if fermented to the same extent [yes, to the same *degree*!] would give a mighty 12% ABV.) Significant quantities of sugar were now being used, which would facilitate these higher gravities. By 1880 the average original extract was 14.25°P, and in 1905 it was 13.25°P. There were some legendary brewing names in the British Isles, immortalized in the verse of C. S. Calverley:

> O Beer! O Hodgson, Guinness, Allsopp, Bass!
> Names that should be on every infant's tongue!

World War I highlighted concerns about excessive drinking. David Lloyd George ranted that "Drink is doing us more damage in the war than all the German submarines put together."

War also has implications for technical issues. For instance, in the Second World War the public wanted volume and were prepared to compromise on strength, and restrictions on the availability of raw materials therefore meant that beer became weaker, the average original gravity now being 10°P. All sides suffered—and brewing of beer in Germany was stopped by decree of the Nazi government on March 15, 1943. The effect on morale must have been substantial. Strong voices in the U.K. government wanted a ban on alcohol, to divert raw materials to food production. However, a calculation showed that if the beer supply was cut in half and the barley saved were diverted to chickens, the net benefit would have been one egg per month in the ration—and severe public discontent.

The U.S. government offered this advice in a booklet for their servicemen stationed in the United Kingdom:

> The usual British drink is beer, which is not an imitation of German beer, but ale. The British...can hold it. Beer is now below peacetime strength, but can still make a man's tongue wag.

Such advice ignored the fact that British ale tends to be relatively low in alcohol, if substantial in flavor. One thing that the North American servicemen will certainly have noticed was the low carbonation of the local ales. The Canadian servicemen in the United Kingdom added salt to their beer, claiming it gave "sparkle" and a good appetizing head.

GERMANIC ROOTS

The western brewing industry first became established in the regions of Bohemia (now part of the Czech Republic) and Bavaria. There were brewers at the court of Charlemagne, who, like Henry VI of England, insisted upon wholesome technology in the production of beer. Up to 500 monasteries at the time were brewing beer, especially the Benedictines.

The importance of good malt to good beer was realized, leading to the development of specialist malt houses. Despite the appreciation of hops coming earlier in Germany than it did in England, there were still plenty of adherents to the merits of gruit (the proprietary blend of herbs and spices used to flavor ale), including the Archbishop of Cologne, who had something of a monopoly concerning the concoction.

It is a myth that lager-style products have always been the "type" beer of Germany. Until the sixteenth century (and not terribly long before the Pilgrim Fathers made their way from England to the New World), ale was the major beer type in Germany. Bottom fermentation probably started in Bavarian monasteries, and was first mentioned in minutes of the Munich town council in 1420. One of the main driving forces for the development of this beer style was an edict of Prince Maximilian I in 1533 that basically precluded brewing in the summer without a special dispensation. The ale style products from top fermentation had been brewed in those warmer summer months, but now the emphasis shifted to the bottom fermentation practices used in the winter, producing beer in sufficient quantities to store (*lager* is from German *Lager* "storehouse") until the subsequent fall, when brewing could start again.

Perhaps the most durable edict was that of 1516 in Bavaria, when Dukes Wilhelm IV and Ludwig X declared the Reinheitsgebot in an attempt to ensure that undesirable materials did not find their way into the brew and to prevent price competition with bakers for wheat and rye. The law survives to this day, extended throughout Germany for domestic brews, restricting the raw materials to barley malt, hops, yeast, and water. As originally conceived, of course, the law did not include yeast, as it hadn't been discovered yet (and read on to see what a prominent German chemist was saying about the existence of that beast as recently as the early mid-nineteenth century).

A BRAVE NEW WORLD

It was the English who brought beer to North America—a reasonable trade-off for potatoes, it might be said. Sir Walter Raleigh is said to have malted corn in Virginia in 1587 (and in South America malted corn had been fermented by the Incas many years before Spanish settlers founded a brewery near Mexico

City in the mid-sixteenth century), but it was the Pilgrim Fathers in December 1620 who shipped the first beer into the country. And why did they land at Plymouth Rock? Because

> We could not now take time for further search or consideration, our victuals being much spent, especially our beer, and it being now the 19th of December.

In fact, the passengers were urged to take to the shore rapidly, so as to leave what remaining ale there was for the sailors.

Adriaen Block, a Dutchman, opened the first brewery in North America, in 1613. It was little more than a log hut in New Amsterdam (which would become New York City). The Dutch West India Company opened the first public brewery in the United States in Lower Manhattan in 1632, with a grist largely of oats. Although the early immigrants were of a somewhat puritanical persuasion, beer was considered (as it still should be) a drink of moderation, and certainly a preferable alternative to the dubious alternatives then available made by the distillation of fermented corn.

It has been claimed that advertisements were soon being placed in London newspapers inviting experienced brewers to immigrate to America. And the first paved street in America—Brouwer Straat ("Brewer Street")—was laid in New Amsterdam in 1657 apparently because one of the brewers' wives despaired of keeping her house clean because previously the thoroughfare had been too dusty. In 1664 King Charles II seized the former Dutch territory of New Amsterdam and set it in the charge of his brother the Duke of York, who certainly had the right idea about properly trained brewers (a legacy remaining to this day, for which I am truly grateful!). The Duke's Laws required that brewing should be carried out by people trained and qualified so to do.

The Scottish and Irish immigrants brought with them a passion for whisky, which in due course overtook beer as the alcoholic beverage of choice, so much so that just prior to the Civil War beer was accounting for not much more than 10% of all the alcohol consumed in this nation.

By the eighteenth century, New York and Philadelphia were the principal seats of brewing, and at the turn of the next century, there were over 150 breweries in the United States, with one-third of them in each of the above two cities. Production, though, was less than 230,000 barrels (U.S.). George Washington had recently died (not before having called for a banning of imports of beer from England to further help the cause of untaxed local brews), leaving his own brewery on the family's Mount Vernon estate in Virginia. Earlier, in the War of Independence, American troops each received a quart (two pints) of beer per day. For that luxury the soldiers had perhaps to thank Samuel Adams, the Massachusetts-based leader of the early independence movement, who was himself a brewer. Boston *Tea* Party, indeed! There is a challenged legend that Thomas Jefferson composed the Declaration of Independence at the Indian Queen Tavern in Philadelphia.

We must move on to the early to mid-nineteenth century, though, to find the beginnings of the great brewing dynasties of the States (table 1.7). Their origins were in Germany. The year 1829 saw the founding of America's longest-standing brewery in Pottsville, Pennsylvania, by David Yuengling. Frederick Schaefer arrived on these shores in 1838 with a dollar in his pocket but with the drive to start a brewery with his brother Maximilian four years later. They began by buying Sebastian Sommers' small brewhouse on Broadway between 18th and 19th Streets in New York, but by 1849 their brew was popular enough they moved uptown to 51st Street and Fourth (now Park) Avenue. Also in 1838, Alexander Stausz and John Klein in Alexandria, Virginia, pursued the first commercial production of lager-style beer in America. Soon most of the urban developments sported their own lager breweries. Some great names emerged (see table 1.7). In 1844, Jacob Best founded the company that would become Pabst, thanks to the wedding of Best's daughter to a steamboat captain named Frederick Pabst. Bernhard Stroh, from a Rhineland family with two centuries of brewing pedigree, opened his brewery in Detroit in 1850. Five years later, Frederick Miller bought out Jacob Best's sons' Plank Road Brewery in Milwaukee (see box 1.5). In 1860, Eberhard Anheuser purchased a struggling St. Louis brewery and, after his daughter married a supplier named Adolphus Busch, an émigré from Mainz, the mighty Anheuser-Busch Company was born (see box 1.4). And a dozen years later, another migrant from the Rhineland, Adolph Coors, set up shop in Colorado (see box 1.6).

By 1873 there were over 4,000 breweries in the United States, outputs averaging some 2,800 barrels each. In all countries, brewing undergoes

TABLE 1.7. TEN LARGEST BREWING COMPANIES IN THE UNITED STATES, 1895

Pabst	Milwaukee
Anheuser-Busch	St Louis
Joseph Schlitz	Milwaukee
George Ehret	New York
Ballantine	Newark, N.J.
Bernheimer & Schmid	New York
Val Blatz	Milwaukee
William J. Lemp	St Louis
Conrad Seipp	Chicago
Frank Jones	Portsmouth, N.H.

Courtesy: Dr W. J. Vollmar

BOX 1.4 ANHEUSER-BUSCH

The story of the leading American brewer and maker of the two largest-selling beers in the world begins in 1860 when 55-year-old Eberhard Anheuser acquired the Bavarian Brewery in St. Louis from George Schneider. There were 40 breweries in the great Missouri city at the time, and Anheuser's ranked 29th. A year later, Eberhard's daughter Lilly married German-born Adolphus Busch, born in 1839 as second youngest of 22 children, who, three years thereafter, joined his father-in-law's company.

By 1865, output was 8,000 barrels a year. In 1876 a new brand was developed by Busch with the assistance of his friend, St. Louis winemaker and restaurateur Carl Conrad, a beer that would satisfy all beer drinkers and be of uniform quality wherever it was consumed. This new beer was called Budweiser. Initially, Busch brewed the product and Conrad bottled and distributed it. In the first year, over 225,000 bottles were sold, and within four years volume had reached 2.3 million bottles. It would, of course, go on to become one of the world's largest-selling beers.

Upon Eberhard Anheuser's death in 1880, Adolphus Busch became president of Anheuser-Busch. He didn't give business acquaintances his card, but rather a pocketknife with a peephole revealing his photograph. Adolphus Busch was the first brewer to recognize how American society was changing and would continue to be changed by advances in technology. The company pioneered the application of the latest developments, such as pasteurization and artificial refrigeration (in the brewing process and also in railcars used to transport the product), which allowed a huge increase in output and the opening up of a vast hinterland for the beer, wherever the expanding railroads were headed. In 1896, he established another great brand, Michelob, and by 1901 Anheuser-Busch was brewing more than 1 million barrels of beer per annum.

Before Prohibition, Anheuser-Busch was shipping its products around the world, and yet the company had foreseen the likely imposition of abstinence and developed a non-alcoholic beverage called Bevo in 1916.

Adolphus Busch died in 1913, the presidency passing to August Busch Sr. (born December 29, 1865), and it fell upon him to guide the company through the years of Prohibition. They made truck bodies, refrigerated cabinets, barley malt syrup, ice cream, ginger ale, root beer, chocolate- and grape-flavored beverages, corn syrup, and baker's yeast. One well-known product was Bevo, which, prior to national Prohibition, hit its sales peak in 1918 with over 5 million cases sold. Upon the repeal of Prohibition, a hitch team of Clydesdales was presented to August Sr., and these animals have been integrally associated with the company and Budweiser ever since. Adolphus Busch III took the company reins in 1934, and his tenure coincided with another great challenge, the Second World War. War bond purchases by Anheuser-Busch employees footed the bill for two B-17 bombers, one of which was called the Buschwacker! The West Coast market was voluntarily sacrificed in order to free trains for military shipments. Nonetheless, Adolphus III

(continued)

BOX 1.4 CONTINUED

presided over a tripling of the company's beer output, at least in part through the advent of canning the product. One year after the end of the war, the company presidency passed to August Busch Jr. (born March 28, 1899).

In his case, the magnitude of company growth was phenomenal, with volume going from 3 to 34 million barrels during his tenure as president. Apart from his enthusiasm for the Clydesdales and the Cardinals baseball team operating out of Busch Stadium in St. Louis, August Jr. also diversified the company into theme parks, with Busch Gardens. His son August A. Busch III (born June 16, 1937) was president from 1975 until 2002, driving the company to unprecedented heights as one of the world's leading brewers and probably the most quality-conscious, with brewing operations across the globe. Among the product triumphs has been Bud Light, launched in 1982 and now the world's best-selling beer. August A. Busch IV took over the reins in 2006, so the great brewing family tradition continues. Late in 2008, as this book headed to press, Anheuser-Busch was acquired by InBev.

In the United States, Anheuser-Busch has breweries in Fairfield (Cal.), Los Angeles (Cal.), Fort Collins (Colo.), Houston (Tex.), St Louis (Mo.), Columbus (Ohio), Cartersville (Ga.), Jacksonville (Fla.), Baldwinsville (N.Y.), Merrimack (N.H.), Newark (N.J.), and Williamsburg (Va.). Their brands are brewed the world over, for they own breweries in countries such as England and China and have their beer brewed elsewhere under license to the strictest of standards.

Left: August Busch III; right: August Busch IV (courtesy of Anheuser-Busch)

The Anheuser-Busch St. Louis brewery (courtesy of Anheuser-Busch)

rationalization, so by the end of World War One there were half as many brew-eries, each producing on average twenty times more beer than 45 years earlier (table 1.8). By the time the Second World War had run its course, there were just 465 breweries in the United States, their output averaging some 190,000 barrels. Compare those volumes with the output of the gigantic Coors brewery in Golden, Colorado, which now produces well over 20 million barrels of beer each year.

The production of lager (a style that the likes of Busch, Miller, Stroh, and Coors would have been more familiar with in their homeland) demanded ice. Accordingly, such beer had to be brewed in winter for storage (lagering) until the greater summer demand. Such protocols were possible in Milwaukee, Wisconsin, using the ice from Lake Michigan and local caves for storing the beer. Milwaukee rapidly emerged as the great brewing center of the States, with Pabst and Schlitz amongst those competing with Miller. Once Carl Von Linde in 1870 showed how machines could be developed to produce ice, lager could be brewed any time—and anywhere. And the application of Pasteur's proposals

BOX 1.5 FREDERICK MILLER

Frederick Edward John Miller was born on November 24, 1824, into a wealthy family from Riedlingen in Germany. For seven years, from the age of fourteen, he studied in France, and prior to returning to Germany, he visited an uncle in Nancy. That uncle, fortuitously, was a brewer, and Frederick liked what he saw so much that he decided to stay and learn the trade. Soon he was in a position to brew his own beers, so he leased the royal brewery in Sigmaringen, back in his home country. With the Germanic Confederation of states in some turmoil, Miller became one of many to seek a new life in the United States of America, where he arrived in 1854 with his young wife Josephine and their infant son Joseph Edward. They had in their possession $9,000 worth of gold.

The Millers spent a year in New York before settling in Milwaukee. Before long, Frederick had bought the Plank Road Brewery from Frederick Charles Best (from the family that developed the Pabst brand) for $8,000. Beer at the time retailed at less than 5 cents a glass in the taverns of Milwaukee. In the first year after he bought the brewery, it produced 300 barrels of lager-style beer. By the time he died in 1888, the annual production was 80,000 barrels.

Miller clearly knew his business. The brewery in the Menomonee Valley had a good water supply and had ready access to excellent barley grown locally. Frederick Miller was a kindly employer, opening a boardinghouse next to the brewery for unmarried staff and, in addition to the free meals (four per day) and lodging, paying them salaries of up to $1,300 a year. They had to work, though, with just a one-hour break in days that started at 4 A.M. and finished at 6 P.M.

Sadly, for such a generous man, Frederick Miller had quite a tragic domestic life. Josephine died in April 1860, having born six children, most of whom did not survive beyond infancy. Miller married Lisette Gross the same year, and they had many children, of whom five survived beyond their fledgling years. It was these children that carried forward the name of the Miller Brewing Company, notable among them being Frederick C. Miller, the grandson of the founder and a Notre Dame football star in his college days. In 1954, Miller was the ninth biggest brewing company in the United States, with production of 2 million barrels. In 1969 Philip Morris Co. acquired a 53% controlling share in the company, buying the remaining shares a year later. In 20 years production increased eightfold, making Miller Brewing Company today the second largest Brewer in the United States, a situation retrenched in 1999 when they and Pabst each acquired parts of the Stroh brewing empire, the latter company sadly exiting brewing after some 150 years. The relentless march of Brewer internationalization found Miller acquired by South African Breweries in May 2002 for $5 billion.

BOX 1.6 MOLSON-COORS

Adolph Coors founded his brewery in the foothills of the Rocky Mountains at Golden, some 20 miles west of Denver, in 1873, and, like Anheuser-Busch and Stroh, the company is still characterized by strong family involvement. For long enough Coors held a certain mystique in some states where it was unavailable, because until comparatively recently Coors beer was shipped to only 11 Western states.

The Coors operation differs from that of the other big Brewers in the United States in that it is concentrated on just two sites, the enormous Golden plant at which Adolph Coors first brewed 124 years ago and, since only 1990, a brewery at Memphis, Tennessee. For ten years Coors has packaged product at Elkton in the Shenandoah Valley in Virginia, the beer being shipped there from Golden in refrigerated tanks on railcars. In the past few years, Coors has moved out of a substantial interest in a brewery in South Korea, but has acquired a hefty slice of the old Bass brewing empire, together with the biggest-selling brand in the United Kingdom, Carling Black Label. (Incidentally, if you travel widely you will find that this brand is very different in the United Kingdom and in South Africa. The brand originated in Canada, but the rights to it are owned separately by Coors and by South African Breweries. The latter company has very much changed the recipe from the original.)

Another unique feature of Coors has been its vertical integration. Although relaxing this to a certain extent recently, Coors has long been substantially self-contained in breeding its own barleys, supplying its own malt, making its own cans, supplying its own energy resources, and so on.

Now, Coors is merged with Molson. As I write, they are still debating where to locate the headquarters: it seems that it can't be Golden or Toronto, where Molson has long since had its corporate center. In the United States there has been a mammoth investment in Shenandoah, with much of the brewing capacity to be located there to satisfy the eastern markets. And late in 2008, Coors and Miller combined operations.

for heat-treating beer to kill off spoilage organisms and the advent of bottle and stopper technology meant that beer could be packaged for home consumption and consumed almost any place, after shipment nationwide on the burgeoning rail network in railcars that were developed with the latest refrigeration technology. Such developments, and also the advent of cans, with their lighter weight as compared to bottles, and metal kegs, which allowed for more robust shipping of draught products, reinforced the hand of the major brewers as they took their merchandise to the great cities across the nation. The

TABLE 1.8. THE CHANGING SHAPE OF THE UNITED STATES BREWING INDUSTRY (MILLION BARRELS IN U.S.)

Year	Anheuser-Busch	Miller	Coors	Schlitz	Pabst	Misc.
1940	2.5	0.5	0.1	1.6	1.6	Schaefer 1.4, Falstaff 0.6, Schmidt 0.6, Stroh 0.5
1950	4.9	2.1	0.7	5.1	3.8	Schaefer 2.8, Falstaff 2.3, Schmidt 1.1, Olympia 0.6, Stroh 0.5
1960	8.5	2.4	1.9	5.7	4.7	Falstaff 4.9, Carling 4.8, Schaefer 3.2, Stroh 2.1, Schmidt 1.8, Olympia 1.5, Genesee 0.8, Heileman 0.6, Pearl 0.5
1970	22.2	5.2	7.3	15.1	10.5	Schaefer 5.7, Falstaff 5.4, Carling 4.8, Olympia 3.4, Stroh 3.3, Heileman 3.0, Schmidt 3.0, Pearl 1.8, Genesee 1.5
1975	35.2	12.9	11.9	23.3	15.7	Schaefer 5.9, Olympia 5.6, Stroh 5.1, Falstaff 5.0, Heileman 4.5, Carling 4.1, Schmidt 3.3, Genesee 2.2, Pearl 1.4, Rainier 0.9, Blitz-Weinhard 0.8
1980	50.2	37.3	13.8	15.0	15.1	Heileman 13.3, Stroh 6.2, Olympia 6.1, Falstaff 3.9, Schmidt 3.6, Genesee 3.6, Schaefer 3.6
1985	68.0	37.1	14.7		11.5 (S & P)	Stroh 23.2, Heileman 16.5, Genesee 3.0
1990	84.6	46.2	19.2		8.2 (S & P)	Stroh 16.1, Heileman 11.2, Genesee 2.2, Gambrinus 0.7, Boston 0.1
1995	84.8	47.7	18.7		?	Genesee 1.8, Gambrinus 1.5, Boston 1.0
1999	95.1	44.0	20.1		11.6*	Genesee 1.3, Gambrinus 3.3, Boston 1.1
2005	118.6	47.4	26.4+			

Value rounded up to first decimal point

S & P incorporated Pabst and Falstaff

* includes Stroh volume

+ Molson Coors

American taste rapidly swung toward the pale, brilliantly clear, relatively dry, and delicately flavored products that are now the norm and that represent two-thirds of beer sales in the United States. The top brands in the United States are listed in table 1.9. The top regional brews are listed in table 1.10: cumulatively they come nowhere near the volume of the major lagers, but they are steadily growing in volume, as we have seen, and are welcomed by brewers and customers alike for the diversity that they introduce. Table 1.11 gives the top beer imports into the United States, while table 1.12 lists the top beer brands worldwide. It is a case of hold onto your hats, however. The word is that as I write (March 2008) a Chinese beer that most of the world has never heard of (Snow) is about to move into second place.

TABLE 1.9. TOP U.S. BEER BRANDS 2007

1. Bud Light
2. Miller Lite
3. Budweiser
4. Coors Light
5. Corona Extra
6. Heineken
7. Natural Light
8. Michelob Ultra Light
9. Busch Light
10. Miller High Life

TABLE 1.10. TOP U.S. CRAFT AND REGIONAL BEERS

1. Sierra Nevada Pale Ale
2. Samuel Adams Boston Lager
3. Blue Moon White
4. Samuel Adams Seasonal
5. New Belgium Fat Tire
6. Samuel Adams Light
7. Shiner Bock
8. Widmer Hefeweizen
9. Samuel Adams Brewmaster's Collection
10. Redhook ESB
11. Pyramid Hefeweizen
12. Deschutes Mirror Pond Pale Ale
13. Redhook IPA
14. Alaskan Amber
15. Deschutes Black Butte Porter

TABLE 1.11. TOP FIVE IMPORTED BEER BRANDS 2005

Brand	Country of origin	Market share of import segment (%)
Corona Extra	Mexico	26.6
Heineken	Netherlands	16.4
Corona Light	Mexico	6.6
Tecate	Mexico	4.7
Labatt Blue	Canada	3.3

From Beverage Industry *State of the Industry Report, 2005*

TABLE 1.12. THE WORLD'S BIGGEST BEER BRANDS (2006)

Brand	Volume (million hectoliters)
Budweiser and Bud Light	90.5
Skol	32.8
Corona	31.8
Snow	30.5
Brahma	25.0
Heineken	24.7
Coors and Coors Light	23.6
Miller Lite	21.6
Asahi Super Dry	17.5
Tsingtao	16.9
Yanjing	15.5
Jinxing	14.7
Polar	14.1
Busch	13.7
Harbin	13.7
Natural	13.3
Zhujiang	12.5
Antarctica	12.4
Carling Black Label	12.1
Carlsberg	11.7

A BRIEF HISTORY OF BREWING SCIENCE

By the end of the seventeenth century, only one textbook on brewing had been produced, by Thomas Tryon in 1691: *A New Art of Brewing Beer, Ale, and Other Sorts of Liquors, So as to Render Them More Healthful to the Body…To which is Added the Art of Making Mault…Recommended to All Brewers, Gentlemen and others that brew their own Drink.* Many years would elapse before the science emerged, slowly at first, which would explain what was happening in the malting and brewing processes and how they could be modified and controlled to

BOX 1.7 BREWING ORGANIZATIONS IN THE UNITED STATES

Brewers worldwide have long since organized themselves into entities designed to ensure excellence in all aspects of beer and the brewing process. In the United States, there have been several of these.

The *United States Brewers Association* was founded in 1862, with some 37 brewing companies represented at the first meeting on August 21, 1862, in Pythagoras Hall in New York. A first convention was held in the same venue in November of the same year. It wasn't until the fourth convention in October 1864 that the name United States Brewers Association (USBA) was adopted; they previously referred to themselves as the Lager-Beer Brewers Association. The types of key issues discussed were taxation and prohibitionist tendencies.

At the ninth convention in Newark, N.J., came discussions about the founding of a brewers' school. Meanwhile, another key topic of debate concerned health and wholesomeness, and at the 17th convention in Milwaukee came a resolution demanding expulsion of any member proved to have added noxious drugs to the beer and "denouncing as false the charges publicly made that brewers were using deleterious substances."

The USBA had a scientific committee, which submitted a report at the 42nd convention in Saratoga, N.Y., containing instructions for the taking of samples and submitting them for analysis. The rules covered "antiseptics, barley, beer, ale, boiler compounds and scale, corn products, filter material, mash-tub grains, hops, malt, oils and lubricants, pitch, rice, sugar, glucose and syrups, varnish, water, wort and yeast."

A year later, in Niagara Falls, N.Y., a resolution was passed reaffirming that brewers would "assist to the full extent of their ability the efforts to promote the objects of the Pure Food Congress." A correspondence was under way with the U.S. Bureau of Chemistry regarding definitions and standards for beer to be fixed by the Association of Official Agricultural Chemists.

On January 1, 1907, the Pure Food and Drug Act went into effect, and this clearly stated the case for all-malt beers. Dialog with the industry, however, led to the acceptance of adjunct usage.

A year later, at the Milwaukee Convention, USBA president Julius Liebmann extolled the merits of beer over other drinks: "…of all alcoholic beverages beer is the mildest, averaging only about three and one-half per cent of alcohol." Meanwhile, the convention was highlighting the Pure Food and Drug Act and the role of the U.S. Department of Agriculture in improving American barley.

The 56th convention was in Cleveland on November 18, 1916, under the presidency of Gustave Pabst. Within the year the U.S. Congress had adopted the resolution for the 18th amendment to the Constitution. And so the 57th convention was not until September 27, 1933, in Chicago, with Jacob Ruppert presiding.

(continued)

BOX 1.7 CONTINUED

The *U.S. Brewmasters Association* was formed in Chicago on March 21, 1887. Its manifesto was "to make the interests of the brewing industry its own." Previously, there had been separate brewmaster societies, inter alia in New York, Cleveland, and Cincinnati. The organization was incorporated in 1912 as the Master Brewers' Association of the United States, morphing to the *Master Brewers Association of America* (MBAA) in 1933.

The goals established for the U.S. Brewmasters Association were "to cultivate mutual acquaintance; to further the interests of the brewing industry; to discuss technical questions relating to brewing; to promote the education of brewery workmen; and to establish a system of apprenticeship." The president of the Chicago Brewers' Association declared at the first meeting, "...we must take active steps to come to the aid of science, because where theory combines itself with practice, there we may be sure that gain to mankind will result." The organization was to have its base in Chicago, and the business language was to be German.

By the time of the second convention in New York in 1888, there were 229 members, and it was recommended that apprenticeships to the industry should be two years in duration and that any company with more than 15 workers should have one apprentice.

Ten years later there were 352 members, and the expressed desire was that, following an apprenticeship, trainees should be directed to a brewers' college. At the 12th convention in Buffalo in 1900, it was declared that apprentices should be 16–20 years old, and should serve two years in a brewery and three months in a malt house. At this meeting, addresses on scientific subjects were delivered by several brewing scientists. The organization periodically published the *Brewers' Calendar*, a reference book containing technical information.

It was not until 1948 that the MBAA created the MBAA Research Foundation in order to pursue and coordinate scientific research in brewing. They placed $50,000 in escrow to establish the capital sum. On Jan. 13, 1950, there was a joint meeting in Chicago of the USBA, the Small Brewers Association (later called the *Brewers Association of America*), and a committee of the MBAA Research Foundation. The topic of debate was a plan for an industrywide research program. After much discussion, there was a revocation of the MBAA Foundation, to be replaced by a much broader *Brewing Industries Research Institute*, founded January 9, 1952, with charter members the American Society of Brewing Chemists, the Barley and Malt Institute, the Brewers Association of America, the USBA, and the MBAA. A later enrollee was the Dominion Brewers Association, which would become the Brewers Association of Canada.

A major problem concerned how to select projects and the issue of publishing the findings. This in turn led to financial pressures, and on January 31, 1969, it was resolved to dissolve the Institute. All surviving assets passed to the MBAA. The parting contribution was a book edited by W. D. McFarlane, the scientific director of BIRI, which summarized and explained the priorities for future research, wherever it may be undertaken. The overwhelming message concerned the need for much more study of beer stability, especially staling.

The pursuit of standardized methods for the U.S. brewing industry commenced with the Analysis Committee of the USBA prior to Prohibition. After repeal, the driving force for the restoration of activity in this area was Max Henius. In the fall of 1934, there was an initial meeting on the topic between the MBAA and USBA, followed by a meeting at the Schwarz labs on October 12 of the same year at which it was agreed to have an initial focus on malt analysis. Six subcommittees were charged with the task of recommending different procedures. Inaugural funding was $250 each from brewers, maltsters, and scientific stations. Soon the committee was considering methods beyond malt analysis, and on October 11, 1935, it adopted the name *American Society of Brewing Chemists* (ASBC). The organization was headquartered at the Wahl-Henius Institute in Chicago and had 24 companies comprising its membership.

It was soon proposed to throw ASBC open to individual membership with a mechanism for submitting research papers. At the meeting in Cleveland on June 17, 1938, the bylaws were adopted establishing the structure of ASBC as it basically is today. The Technical Committee was formed to continue the work on methods of analysis. Leo Wallerstein became chair of the membership committee, and within four months there were 31 members and 65 applications pending. The society was positioned to further scientific research in brewing, and the first meeting at which original research papers (thirteen in toto) were offered was in Kansas City, Mo., from May 22–24, 1939. The audience comprised 33 members (approximately one-third of the total roll). The year 1940 saw the first publication of the *Proceedings of the American Society of Brewing Chemists*, which became the *Journal of the American Society of Brewing Chemists* in 1976. (The author of this book has been its editor-in-chief since 2000.)

Liaison with the MBAA commenced in October 1945, with an initial focus on "The keeping qualities of beer." By 1963 there was talk of a merger between ASBC and MBAA, and the issue was put to the vote two years later but did not pass. In 1977, a joint planning committee worked to arrange a scientific meeting encompassing both organizations, and as a result the first World Brewing Congress was held in St Louis, Mo., in September 1984.

A network of local groupings within the ASBC developed from initial gatherings in Cincinnati and Philadelphia in 1952. Local sections were formally recognized in 1966, with New York being the first such assembly.

It was in April 1880 that the Regents of the University of California were mandated by the California state legislature to establish a program of instruction and research in viticulture and enology, reflecting the clear potential for California to develop an

(*continued*)

BOX 1.7 CONTINUED

international wine business. Two years after the repeal of Prohibition in 1933, the department became established on the campus at Davis. In 1956 Ruben Schneider, technical director of San Francisco's Lucky Lager Brewing Company, wrote to Emil Mrak, then chair of the Department of Food Technology at UC Davis (later Chancellor), urging the establishment of a brewing program to complement the wine focus. The brewing program was finally established in 1958 with a grant from the MBAA, Schneider now being its president. The brewing technology course was the first of its kind to be offered in a major American educational institution, and the program continues to this day.

Breeding and associated research programs on hops have been prominent at Oregon State University (OSU, Corvallis) for many years. Research on hops in the United States has been coordinated and funded through the member-based *Hop Research Council* since 1979.

Fundamental attempts to develop improved malting barley varieties in the United States began in 1938 with the founding of the Malt Research Institute (MRI) in Madison, Wisc. The MRI coordinated the evaluation of barley varieties and funded research at the USDA-ARS Cereal Crops Research Unit at the University of Wisconsin. The Midwest Barley Improvement Association was formed in 1945 in Milwaukee, Wisc., and in 1954 expanded to include the whole country, becoming the Malting Barley Improvement Association (MBIA). The MRI was merged with MBIA in 1959, and in turn the *American Malting Barley Association* (AMBA) emerged from these forerunners in 1982. This body supports fundamental and applied research through the awarding of grants.

Nowadays the in-house research effort pursued by production brewing companies is very limited. However, prominent programs have been pursued through the years by all the major players, and until relatively recently, a significant proportion of their research was published in the pages of the *MBAA Technical Quarterly* and the *Journal of the American Society of Brewing Chemists*. Competitive pressures nowadays dictate that these companies rarely if ever publish original research findings; indeed, the programs of brewing companies in the United States are now very much focused on competitive advantage.

ensure the production of consistent products of high quality. It is this science, and the refined technology that developed from it, which forms the heart of this book.

It was in 1680 that a 47-year-old draper from Delft in Holland, Antonie van Leeuwenhoek, reported to the Royal Society in London how he had developed a microscope that had enabled him to inspect a drop of fermenting beer and reveal therein something we now recognize as yeast cells. One hundred and

BOX 1.8 HAPPOSHU AND THIRD CATEGORY PRODUCTS

The Japanese brewers discovered a loophole in the beer taxation laws which showed that products containing less than 25% malt would attract substantially less duty than "conventional" beers. As a result, *happoshu* was born: products with a sufficiently low level of malt, with the balance replaced by adjunct. The products could not be labeled "beer," but in every respect they clearly emerged from the same stable: bottle, label, location on the supermarket shelves, alcohol content, foam, color, and so on. The flavor clearly was not as good as "regular" beer, but for day-to-day consumption in the privacy of one's own home, it made sense to buy them, with the premium beers purchased when there were visitors to impress.

The Japanese authorities identified the loophole and tightened matters up. However, a loophole remained: if a product contained zero malt, then the rate of taxation was lower still. Thus emerged "Third Category" products, based on diverse adjuncts (of the type I cover in chapter 8) but also the likes of soybean and pea. They are not to my palate—but what do I know?

---

fifty years later Charles Cagniard de Latour in France and Theodor Schwann and Friedrich Kützing in Germany independently claimed that yeast was a living organism that could bud. They were ridiculed by the Germans Justus von Liebig and Friedrich Wöhler, who insisted (we believe with sarcasm) that yeasts comprise eggs that turned into little animals when put into sugar solution. Liebig and Wöhler, who clearly had little sympathy with matters biological, suggested that these animals

> have a stomach and an intestinal canal, and their urinary organs can be readily distinguished. The moment these animals are hatched they begin to devour the sugar in the solution, which can be readily seen entering their stomachs. It is then immediately digested, and the digested product can be recognized with certainty in the excreta from the alimentary canal. In a word, these infusoria eat sugar, excrete alcohol from their intestinal canals, and carbonic acid from their urinary organs. The bladder, when full, is the shape of a champagne bottle, when empty it resembles a little ball; with a little practice an air-bladder can be detected in the interior of these animalculae; this swells up to ten times its size, and is emptied by a sort of screw-like action effected by the agency of a series of ring-shaped muscles situated in its outside.

It was another Frenchman who sorted the matter out. Louis Pasteur (1822–1895) became professor of chemistry at Lille University and was urged by the local brewers to sort out the difficulties they were having with beer going sour after fermentation. He demonstrated that the infection was due to airborne organisms that he could trap in guncotton and that could be inactivated by heat. By 1860, this tanner's son from Dole was able to conclude that "alcoholic fermentation is an act correlated with the life and organization of the yeast cells."

The informed brewing historian Ray Anderson has eloquently described how Pasteur's role, while pivotal in the history of brewing science, was not absolute. As Anderson says:

> Pasteur's genius—and make no mistake he was a genius—was in bringing together disparate elements and making the whole greater than its parts. What sets Pasteur apart is the rigor of his scientific method, the clarity of his vision in recognizing the significance of his results and in applying his findings to practice.

Anderson, then, emphasizes the contribution of those such as Carl Balling, who spoke in the 1840s of the living nature of yeast in his lectures to brewers in Prague. There were James Muspratt and Heinrich Böttinger (the latter head brewer of a brewery in Burton-on-Trent) who disagreed with their teacher Liebig and who also recognized the criticality of live yeast. There was Jean-Antoine Chaptal, a French chemist who, in 1807, associated films of vegetation on wine with souring. In fact, the present shape of the brewing industry as a well-controlled, highly efficient, and reliable multibillion-dollar industry is testimony to the researches of many eminent scientists working not only on yeast but on the germinative properties of barley, the composition of hops, and the refinements of the malting and brewing processes in their entirety.

A seminal moment in the shaping of the modern brewing industry came in 1883. Emil Christian Hansen, head of the Physiological Department of the Carlsberg Laboratory in Copenhagen, proposed that the all too frequent occurrence of brews that produced unsellable product was not necessarily due to infection by bacteria, as Pasteur had proposed, but rather was because of "wild yeasts." The term "wild yeast" persists to this day, and is really a reference to any yeast strain other than the one that the brewer intends should be used to ferment the beer, for it is that yeast that contributes substantially to the unique character of a beer. It was Hansen who perfected a system for purifying yeast into a single, desired strain, and this forms the basis for the brand-to-brand individuality of beers to this day.

In the ensuing 100-plus years, the technology for the malting of barley and brewing of beer has advanced remarkably, building on the scientific explorations of many gifted scientists worldwide. The processes are enormously more efficient now than they were even 50 years ago. For instance, the malting process is now completed in less than a week, whereas it took twice as long half a century ago. Brewing can take as little as 1–2 weeks, although many Brewers insist on longer processing times: Brewers take pride in their products and, while striving for efficiency, won't take short cuts if quality would be jeopardized.

Brewing scientists, too, have bequeathed to society many concepts that are now accepted as commonplace. For instance, James Prescott Joule was employed in a laboratory at his family brewery in Salford, England, when he contemplated the research that led to the First Law of Thermodynamics. Søren Peder Lauritz Sørensen, working in the Carlsberg Laboratory, explained the concept of pH (the universal scale for measuring acidity and alkalinity) and its importance in determining the behavior of living systems, notably through an impact on enzymatic activity. W. S. Gosset, who was breeding new varieties of barley and hops for Guinness, published under the pseudonym "Student," a name familiar to those statisticians everywhere who apply the *t*-test. Not least, of course, the impact of Pasteur on modern society extends far beyond beer.

This chapter has given us a feel for the magnitude of the world beer market, the pressures that come to bear on it and influence production outputs, and how its shape today is a direct reflection of a long-standing pedigree. It's now time for us to understand the essence of the remarkable processes involved in converting barley and hops into the world's favorite alcoholic drink.

GRAIN TO GLASS

THE BASICS OF MALTING AND BREWING

To start our journey through the art and science of brewing, we will begin with an overview of the entire process from barley to beer. In subsequent chapters, the individual stages will be covered in more detail.

The staple ingredients from which most beers are brewed are malted barley, water, hops, and yeast. The nature of beer is a result of these raw materials and of the two separate (but related) processes that have been used to make this drink for thousands of years. In Germany, legislation has decreed that beer production will involve these materials *alone*. Excellent beers are produced in Germany, but so, too, are they produced in the rest of the world, where there has long been greater flexibility in the materials available to the Brewer. The wherewithal to use a selection of adjuncts, for instance, enables the Brewer to provide the consumer with an excellent selection of beers to meet every drinking occasion. The opportunity, too, to use process aids such as clarifying agents and stabilizers ensures the Brewer's capability to produce beer that will have good shelf life in an economic manner, benefits passed on to the consumer.

The Brewer is not unrestricted in what can be used: in all countries legislation dictates what may be legitimately employed in making beer, what the label has to declare, and how beer may be advertised. In some countries, such as the U.K., the package must give information concerning the date before which a beer should be consumed. In some countries, Brewers must provide ingredients labeling on the container. In the United States, the Alcohol and Tobacco Tax and Trade Bureau (TTB) within the Treasury Department regulates all aspects of the alcohol industry. There is no requirement to list ingredients, other than sulfur dioxide if its level exceeds 10 parts per million (10 mg/L). You will be hard-pressed in the States to find this written on a label, because most brewers strive to ensure that the sulfite level is well below this level. Winemakers generally use sulfites to preserve color and taste, hence "Contains sulfites" is usually found on the label.

At the simplest level, malting and brewing represent the conversion of the starch of barley into alcohol. Brewers are interested in achieving this with maximum efficiency in terms of the highest possible alcohol yield per unit of starch. At the same time, though, they insist on consistency in all other attributes of their product—foam, clarity, color, and, of course, flavor.

When we speak of barley in a brewing context we are primarily concerned with its grain, the seeds growing on the ear in the field (figure 2.1): it is these that are used to make beer. Barley grains (figure 2.2) are hard and difficult to mill. Try chewing them if you will—but have a good dentist on hand! They also don't taste particularly pleasant, drying the mouth and leaving a powdery and astringent aftertaste. Indeed, beer brewed from raw barley is not only troublesome in processing, but it also has a definite grainy character. It must have been pure serendipity when the process of malting was discovered some 100 centuries ago, but upon such happenstance has sprung up a mighty industry responsible for converting this rather unpleasant cereal into a satisfying malt.

FIGURE 2.1 Barley in the field (courtesy of Assured UK Malt)

FIGURE 2.2 A barley kernel (courtesy of Assured UK Malt)

MALTING

Barley is first steeped in water, which enters the grain through the little hole at the embryo end called the micropyle (see figure 2.3). The water then distributes through the food reserve (the starchy endosperm). The first tissue to be hydrated is the embryo, which springs into life. It is the infant plant, and it produces hormones that journey to the tissue (called the aleurone) that immediately surrounds the starchy endosperm. These hormones switch on the production of enzymes, which first chew up the walls of the cells in the aleurone and then move into the starchy endosperm, digesting its walls and some of its protein. As these are the materials that make barley hard, it is this hydrolysis that renders the grain friable, easily chewed, and, subsequently, more readily milled in the brewery. The experienced maltster will evaluate how well this "modification" process is proceeding by rubbing or squeezing individual grains between his fingers. Happily, in the relatively short periods of time needed to soften grain (typically four to six days), less than a tenth of the starch in the endosperm is removed, although the starch-degrading enzymes produced in the aleurone do bore holes in it. The starch is the material that the Brewer will subsequently use as a source of fermentable sugars to make beer: the more of it survives malting, the better.

The cell wall and protein polymers are broken down into small soluble molecules that migrate to the embryo for its nourishment. Using this food,

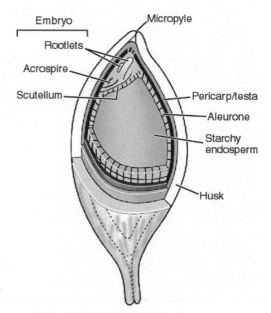

FIGURE 2.3 A diagrammatic section of a barley kernel

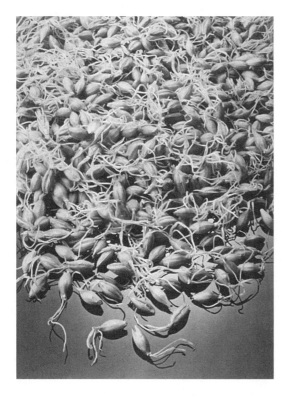

FIGURE 2.4 Sprouted barley
(i.e., "green malt")

the embryo starts to germinate, producing rootlets and a shoot (acrospire)
(figure 2.4). Excessive production of these tissues is not desirable, as this consumes material that can otherwise be sold to the Brewer. The rootlets emerge
through the micropyle and become the first obvious manifestation of germination. The acrospire, of course, heads in the opposite direction to the roots (as
shoots tend to do!) and grows beneath the husk, eventually to appear out of
the distal tip. If the acrospire does appear in a commercial malting operation,
germination has gone too far.

When the germination stage is deemed to have lasted long enough, it is
stopped by heating the grain in a process referred to as kilning. The aim is to
drive off water until the moisture level in grain is below 5%, when the metabolism of the barley will be halted and the product stabilized. The heating process needs to be conducted carefully. If the brewer is to be able to get access
to the starch in the grain, she will need to use the enzymes (the amylases) that
are present in the grain and that are mostly produced during germination.

Enzymes are, for the most part, susceptible to death by heat, and they are particularly sensitive at higher moisture levels. For this reason, the kilning process is started at quite a low temperature (perhaps 50°C, 122°F). When about
half of the water has been removed, the temperature can be raised, and this
ramping will continue according to a preset regimen, depending on the nature
of the malt required.

Malts destined to go into the production of ales are kilned to a higher temperature. This has two implications. The first is that these malts will be darker.
In the kilning process, the breakdown products (amino acids and sugars)
released from proteins and carbohydrates during germination meld to form
so-called melanoidins, which are colored. The higher the temperature (and the
more breakdown products in the first place—i.e., the more extensively modified is the grain), the darker the color.

The second implication of higher kilning temperatures is the development
of complex flavors. The pleasant flavors that we associate with malt and that
enter, for instance, into malty bedtime drinks are produced during the kilning process, also from the interactions between the breakdown products from
protein and carbohydrate. If malt is kilned to particularly high temperatures,
it is possible to make especially dark products (the sort that are used to color
stouts) and to develop flavors described as "burnt" and "smoky."

Malts destined for lager-style beers are generally less extensively modified than those aimed at ale production (i.e., they contain less amino acid and
sugar), and they are kilned to a relatively mild regime. They therefore develop
less color and give quite pale, straw- or amber-colored beers. They may also
deliver a wholly different kind of flavor, one which tends to be more sulfury.

BREWING

It is very unusual for a malthouse and a brewery to be on the same site, even for
so-called brewer-maltsters, which are Brewers that produce their own malt.

The first step in brewing is the milling of the malt (figure 2.5). The phrase
"grist to the mill" is, of course, an accepted part of the English language. Malt
is the principal grist material used for brewing, but there may be others, too,
such as roasted malt or barley, corn, or rice.

All of the unit operations within the brewery must be performed correctly
if the process is to be efficient and trouble-free. Milling is as important as any
stage that follows in the brewery. The aim in milling is to produce a distribution of particles that is best suited to the subsequent processes in the brewhouse. In large part, the malt should be converted to flour with particles small
enough to enable access of water. This will hydrate the particles and enable
the enzymes in the malt to be activated. It will also solvate the substrate

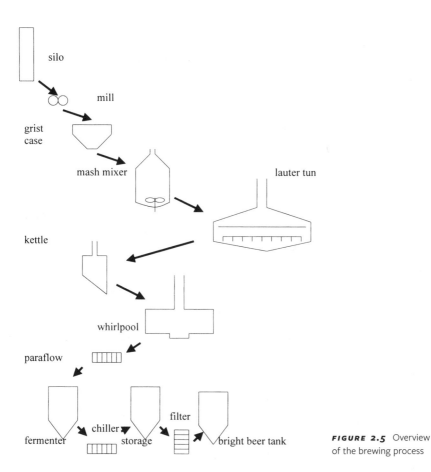

silo

mill

grist case

mash mixer   lauter tun

kettle

whirlpool

paraflow

filter

fermenter   chiller   storage   bright beer tank

FIGURE 2.5 Overview of the brewing process

molecules (principally starch) that the enzymes are targeting. For most brew-houses, though, it is important that the husk component of the malt remain as intact as possible after milling. This is because it will be used to form the filter bed through which the solution of sugars produced in the mashing operation will be recovered in as "bright" a condition as possible.

The milled grist is stored briefly in the grist case before going to the mash mixer (mash tun, conversion vessel). Here it is mixed intimately with warm water to commence the hydrolysis process. Mashing is often commenced at a relatively low temperature (say 45°C–50°C, 113°F–122°F) to enable the more heat-sensitive enzymes to do their job. These include the enzymes that degrade any cell wall polysaccharides surviving the malting process. Then, after perhaps 20 minutes, the temperature is raised to at least 65°C (149°F), for it is at this temperature that the starch is gelatinized. This process can be likened to melting. It involves the conversion of starch from a crystalline, difficult-to-digest structure to a disorganized state readily accessed by the amylase enzymes responsible for chopping it up into fermentable sugars. Happily, the amylases largely survive this higher temperature. The mash is held for a period of perhaps an hour before the temperature is raised once again, this time typically to 76°C (169°F). This serves to stop most enzymatic activity, as well as reducing viscosity and sticking particles together, thereby improving the fluidity of the mash.

In most breweries the sugar solution produced (wort) is separated from the spent grains in a vessel called a lauter tun. The bed depth is relatively shallow, and rakes are used to loosen the bed structure and overcome compacting. Efficient lautering is a skilled operation, the aim being the recovery of as clear a wort as possible ("bright wort") containing as much as possible of the soluble products of mashing (the sum total of which is called "extract"). It is also generally important that the recovered wort is relatively concentrated—so-called high-gravity wort—if production throughput in the subsequent fermentation stage is to be maximized. To facilitate washing of the breakdown products (made from carbohydrate and protein) out of the mash bed, hot water is used to "sparge" the grains. Clearly, too much water must not be used if the wort is not to be excessively diluted. The aim, though, is extraction of as much of the fermentable material as possible from the grains within the restricted time available—the more rapidly the wort can be recovered from the residual grains, the more brews can be performed per day. Almost without exception, the spent grains are sold off as cattle food.

Wort flows directly (usually) from the lauter tun (or one of the other wort separation processes that we will come to in chapter 8) to the kettle (sometimes called the "copper," irrespective of the fact that these days they are mostly fabricated from stainless steel).

Wort boiling, which is performed in this vessel, serves several functions. Foremost among these is the extraction of bittering materials and of aroma components from hops. Traditionally hopping was by adding whole cone hops, and this is still practiced in a good many breweries. The hop residue will still remain after the boil and, analogously to the situation with malt husk and wort separation, the residual hops are used in a so-called hop back to form the filter medium through which the bittered wort is separated. More frequently nowadays hops will have been preprocessed. It is very common for hops to be milled and pelletized before entering the brewery, in which case vegetative matter does not survive intact and the post-kettle vessel is the whirlpool (see below). Alternatively, liquid extracts of hops are used.

Hops contain resins that are extracted in the wort boil and converted into more soluble and bitter forms. Hops also possess a complex mixture of

essential oils, and it is these that provide the different types of hoppy character that can be associated with beers. These molecules are quite volatile and
will evaporate to a greater or lesser extent in the boil. Hops added at the start
of a boil, which typically lasts for one hour, will lose essentially all of their
oils. For this reason, in traditional lager brewing in Europe, a proportion of
the hops is held back for addition during the final few minutes of the boil,
thereby enabling a proportion of the essential oils to survive and provide distinctive aroma notes. This procedure is called "late hopping." In traditional ale
brewing in the U.K., a handful of hops is added to the cask prior to its leaving the brewery. This so-called dry hopping makes for a much more complex
hop character in a beer, as there is no opportunity for evaporation of any of
the oils.

Apart from extraction of substances from hops, wort boiling serves to concentrate wort to a greater or lesser degree (depending on the rate of evaporation allowed, which can range from 4% to 12%), driving off unwanted flavor
molecules, inactivating any enzymes that might have survived mashing and
wort separation, and sterilizing the wort. (Because of boiling, and also because
the antimicrobial bitter compounds are introduced during it, there was a time
when beer was far safer to drink than the local water, which carried diseases
such as cholera and typhoid. It may still be the case in some countries that
beer should be your preferred drink, for this reason.) Most importantly, the
boiling also causes coagulation of much of the protein from the malt, a process
that is promoted by tannin materials extracted from the malt and hops. This
precipitation, to form an insoluble complex called trub (rhymes with "pub" in
England but with "lube" in the States!), is important, as these proteins, if not
removed here, will be capable of dropping from solution in the ensuing beer
to form unsightly hazes and sediments.

In most breweries, the next stage involves the whirlpool, first used by the
Molson company in Canada. The principle stems from Albert Einstein stirring
his cuppa in pre–tea bag days. He noticed that the leaves in the swirling liquid collected at the center of the cup. Eureka! In a brewery the boiling wort is
passed tangentially into a large vessel (the whirlpool, sometimes called a hot
wort receiving vessel) and allowed to swirl there for a period of as short as a
few minutes or as long as an hour. By centripetal forces, the trub collects in
the central cone at the base of the whirlpool, leaving bright wort above it. The
removed trub is often mixed in with the spent grains (and mixed well, because
of its intense bitterness!) before sale for cattle food.

The wort is now almost ready for fermentation—but first it must be cooled
before yeast is added. This is achieved using a heat exchanger, commonly
referred to as a Paraflow, in which the wort is flowed through channels against
a flow of cold water or other coolant in adjacent channels. Heat transfers from

the wort to the water, the latter being recovered for cleaning duties. The wort will have been cooled to the desired temperature for fermentation, which may be as low as 6°C (43°F) for traditional lager brewing in mainland Europe or as high as 15°C–20°C (59°F–68°F) for ale brewing in England.

Prior to addition of yeast, a little oxygen (or air) will be bubbled into the wort. Although the fermentation process leading to the production of alcohol is anaerobic, yeast does require some oxygen, which helps it to make certain parts of its cell membrane and allows it to grow.

The traditional distinction between brewing yeasts divides them into two types, top-fermenting yeast and bottom-fermenting yeast. The first type was traditionally used for ale brewing in open fermenters in the U.K., and such strains have their name because they migrate to the surface of the beer during fermentation. Bottom fermenters, as the name suggests, settle to the base of the fermentation vessel, and they are traditionally associated with the production of lager-style beers. These days the distinction is blurred, insofar as ales and lagers are frequently fermented in the same type of vessel. Although traditional fermenting systems survive, the most common system is the cylindroconical tank, within which the distinction between different flotation characteristics of yeasts becomes distorted.

Fermentation is primarily concerned with the conversion of sugars into alcohol, and the rate at which this occurs is basically in direct proportion to the temperature and to how much yeast is "pitched" into the fermenter. Ale fermentations can be as fast as two or three days, whereas traditional lager fermentations can take more than a fortnight to be completed. The process, however, represents more than simply alcohol production; otherwise, the temperature employed would be substantially greater. Brewery fermentation is also about producing a subtle mix of flavor compounds. The balance of these will depend on the yeast strain involved, which is why Brewers jealously guard and protect their own strains: the character of a beer often depends as much as anything else on the yeast, particularly for the more subtly flavored lagers.

Different shades of opinion govern what happens next. The traditional Brewer of lager beers will insist that a beer must be stored (lagered) on a decreasing temperature regime from 5°C to 0°C (41°F to 32°F) over a period of many months. Others are convinced, however, that no useful changes in beer quality occur in this time and that this period can be substantially curtailed. All are agreed, however, about the merits of chilling beer to introduce stability to it. For most Brewers, this will involve taking the beer to as low a temperature as possible, short of freezing it. In practice this means –1°C or –2°C (30.2°F–28.40°F) for a few (perhaps three) days. Whereas heat-precipitable proteins are removed in the boiling and whirlpool operations, it is the cold-sensitive proteins that drop out at this conditioning stage. The colder the better: –2°C

(28.4°F) for one day is far better than +2°C (36°F) for two weeks. Of course, not all beer is chilled: the traditional English ale, for instance, is clarified using protein preparations known as isinglass finings, which are extracted from the swim bladders of certain types of fish. The isinglass promotes the settling of solid materials from beer.

Once again, the beer needs to be clarified. This can be achieved using various types of filter. Generally, clarification will be assisted by the use of a so-called filter aid, such as kieselguhr, which serves to keep an open bed through which beer can flow, but also to provide pores that will trap solids. Kieselguhr is a diatomaceous earth, a mined substance comprising the skeletons of primitive organisms.

At this stage, too, various materials may be added to promote the stability of the beer. Some of these materials remove the protein or polyphenols that cause hazes. Others are antioxidants that prevent beer from staling. Some Brewers will employ an agent such as propylene glycol alginate, derived from seaweed, to promote foam stability, though there is a strong additive-free policy for most beers in North America.

The beer is filtered into the so-called bright beer tanks, where it awaits packaging. The Brewer will ensure that it has the correct carbon dioxide (CO_2) content. CO_2 is, of course, a natural product of fermentation, but its level in bright beer may have to be increased to meet the specification. Equally, it may have to be lowered: some beers should contain less CO_2 than that which develops in deep fermenting vessels. Some beers have nitrogen gas introduced into them at this stage to enhance foam stability.

Finally the beer is packaged, either into can, bottle (glass or polyethylene terephthalate, known as PET), keg, cask, or bulk tank. The packaging process must be efficient in terms of speed but also quality: there should be no oxygen pickup in the beer, for this will cause the product to go stale. Consistent fill heights must be achieved to satisfy weights and measures legislation, no foreign bodies must enter the package, and, last but not least, the container must be attractive and not damaged during the filling process, which, in the case of cans, might be at rates of over 2,400 cans per minute.

And that's all there is to it! Rigorous selection of raw materials, proper processing, love, and devotion. And opportunities by playing tunes with grist, hops, yeast, process, and more besides, to arrive at the rich diversity of beers that we will contemplate in the next chapter.

EACH TO HER OWN

BEER STYLES

Fundamentally, beers may be divided into ales (including porters and stouts) and lagers.

Traditionally, ales were brewed from relatively well-modified malts that had been kilned to quite high temperatures (lots of color and flavor). They were fermented at relatively warm temperatures (15°C–25°C; 59°F–77°F) by *top-fermenting* yeasts, those that migrate to the surface of the brew in the fermenting vessel. They were produced in open vessels from which the yeast was "skimmed" as a means of collection. Quantities of whole hops were added to the final product to deliver a robust "dry hop character," and they were dispensed at relatively warm temperatures (10°C–20°C; 50°F–68°F).

Lagers, on the other hand, were historically produced from relatively lightly kilned (and therefore very pale-colored) under-modified malts by more complicated mashing regimes. A proportion of the hops were added late in the kettle boil, to allow survival of some hop oils as "late hop character." Lagers were fermented at somewhat cooler temperatures (6°C–15°C; 43°F–59°F) using *bottom-fermenting* yeasts that sedimented during the process and were collected from the base of the vessel. The beers were stored cold for protracted "lagering" periods, and final dispense was cool (0°C–10°C; 32°F–50°F).

The late twentieth and early twenty-first centuries, however, have been marked by considerable blurring of the boundaries that divide these beer styles. The successful brewing companies are characterized by strong new product development programs, from which have emerged some remarkable beers that don't fall easily into any recognized classification. Where, for instance, would you pigeonhole a stout containing oysters or chocolate, ales tasting of heather, or lagers with just a hint of citrus or a whole chili? Even more fundamentally, beers that may fall into an obvious genre in one market may be slotted into an entirely different category elsewhere: for instance, a beer that may be described as a "bitter" in Australia would to an Englishman be perceived as having the characteristics of a lager. I recall a beer brand in England called Long Life that started off its "career" as an ale but, when the English Brewers finally discovered what others had known about for a long while, namely lager, was suddenly reclassified into the sexy new (*sic*) genre.

It is a fact, sad or otherwise depending on your point of view, that it is increasingly difficult to classify beers. This has been exacerbated by the tremendous surge of new product development ideas that has characterized the brewing industry in recent years. Some of these new products have been truly "out of left field." Some years ago, the Miller Brewing Company unveiled a "clear beer" from which all the color had been stripped out; it appeared, through that company's trademark clear glass bottles, to be "water white." It seems that the customer was confused about what the purpose of such a product really was, and it lasted only a very short while. A failure born of technology push over market pull.

Most new products have adhered to established convention in terms of appearance. Modern technology, though, has permitted the extension of the list of beer categories to light beer, ice beer, dry beer, non- or low-alcohol beer—and the opportunities don't end there. A beer is increasingly characterized by either a technological story told about it (e.g., ice beer), an image (e.g., dry beer), or a particular property the consumer expects from it (e.g., light beer or low-alcohol beer). For the most part, though, a beer nowadays seems to be what you choose to call it, and, generally, that still breaks down to ales, lagers, and stouts.

TOP FERMENTATION BEERS

ENGLISH PALE ALE

Referred to as "bitter" when dispensed on tap (draft), these beers can have an alcohol content of anything between 3% and 7.5% by volume (ABV), with a tendency to be at the lower end of this range. In the United Kingdom, these beers sold on draft may either be "cask-conditioned" or "kegged." The latter are "brought into condition" (i.e., carbonated) in the brewery and are filtered to remove yeast and other insoluble materials before being pasteurized prior to filling into kegs. Cask ales are the traditional products of British ale breweries. After fermentation, the ales are "racked" into casks (formerly of wood, but these days usually of aluminum or stainless steel); also introduced are "priming" sugars, a handful of hop cones (for dry hop flavor), and isinglass

BOX 3.1 BEER OR WINE?

I detest styles of advertising in the United States that involve slamming the products of competitors. It is unimaginative and offensive. Regrettably, it has been only too common a format. Several years ago, one craft brewer (that actually at the time brewed all their beer under license, not owning a single brewery of their own) positioned their beer on a reprehensible platform of attacking a notable imported lager. Various other smaller companies have taken pot shots at the big guys. And most recently, we have had the embarrassing clash of two real giants.

The sniping cheapens beer in the eyes of many. I suspect the net effect is to tell many a prospective purchaser that none of this beer stuff can really be any good. How, on the one hand, can somebody say their beer tastes like an angel weeping on your tongue while the other guy says with equal conviction that it stinks (or, just as bad, tastes of nothing)? The consumer is confused and turns elsewhere, to products that they believe they can trust and that are universally championed as the smart thing to be seen with. They embrace drinks, such as wine, that are unanimously championed by all their producers. It's not a case of "ours is great, but theirs is nasty," rather it's "we all make a great drink—we are talking different degrees of excellence."

I am well aware of the fact that the beer commercials can be hilarious. In fact, they are generally screamingly funny. Equally, I am well aware of who the primary purchasers of most beers are—young men who like to chug large volumes and who find risqué ads tremendously funny. However, surely there is a huge untapped market out there that does not switch on to bosoms and belching, those who seek the higher moral territory.

And so a recent Gallup poll in America suggested that wine has overtaken beer as the preferred alcoholic drink in the States. Of those questioned, 39% said that they prefer wine, only 36% beer. Meanwhile, in the United Kingdom, more is now spent per annum on wine than on beer. The official explanations are the lack of a major sporting event (beer does better than wine on these occasions), an increase in the proportion of alcohol being consumed by women, and "clever marketing." I would suggest that the last of these is by no means the least significant.

Winemakers speak of *terroir*, the impact of the land and locale on grape quality and hence wine excellence. There is absolutely no reason why brewers can't make similar overt use of the provenance of their raw materials. They have, of course, done this from time to time, not least mention of refreshing mountain water. But, while there are occasional advertisements that champion the grain or the hops, they are pushed into the background by images of raucous party-going among the young, men dancing with dogs, and football-juggling turtles.

In an era of premixed sweeter beverages (malternatives) and a "let's have fun" attitude, the image of barley waving in the wind or hops looming bold in Kent or Oregon is not prominent in a customer's mind's eye. This is not to say that there is not a vast opportunity for brewers to appeal to those of higher mind, or even the young folk in their moments away from the party scene. For the most part, wine is not a fun drink; rather it is a drink of sophistication, one that stirs passions, debate, and rhetoric and is altogether more cerebral. Beer can do just that, but too often it doesn't.

One has only to venture into the local bookstore to see evidence aplenty. The shelves are replete with volume after volume of quality writing on wine, from Larousse through Sotheby to Oxford. By contrast the literature on beer is sparse—even the best of writing finding it impossible to steer clear of dodgy areas such as listing the numerous synonyms for describing the state of drunkenness. There is no true beery equivalent to *Wine Spectator*.

Wine has even reached the movies, with *Sideways* (supposedly another word that refers to a state of inebriation, though I have never heard it) leading directly to a dramatic upswing in sales of pinot noir in California.

To venture to Napa on a summer's weekend is to encounter droves of traffic trooping up and down the valley from one spectacularly beautiful winery to another, each charging ludicrous sums for sipping thimblefuls of wine in pricey glasses, tempting the would-be connoisseurs to further invest in cases of frequently dubious vintage. Breweries, of course, are seldom places of great loveliness or charm, either in themselves or in their situations. They generally tend to the industrially sophisticated, with conveyor belt throughputs. Wineries, by contrast, are sleepy hollows, leaping into life for the very few weeks of Crush, allowing plenty of time and scope for beautification. Yet surely there is much more that breweries can do to render themselves less the Ugly Sister.

What wool is actually being pulled over eyes by the wine folks—or, viewed from another direction, what opportunities exist for brewers to make a killing on a platform of deterioration? Although I forcefully champion the concept and correctness of freshness in beer, is it impossible to countenance the diametric opposite, and market some beers overtly on their age? From time to time, of course, there have been such products. One that springs to mind is Thomas Hardy Ale. I invite you, however, to consider some of the flavors that are prominent in wines that I am told have genuine excellence, and you will find compounds generally considered to have no place in beer. Diacetyl, for instance. Wandering into the world of fortified wine, think of the *estufagem* process in madeira production: the wine is heated at up to 50°C for as long as 3 months. If you have tasted a beer that has endured that sort of temperature for just a day or two, you will readily appreciate what sort of notes are present. Alcohol content is quite high in madeira, of course. But so can it be in many beers.

Madeira wine, indeed any wine, is seldom consumed straight out of the bottle. Serving wine (usually at exorbitant prices) in a restaurant is a ritual. Smelling the cork, the swirling in the glass, the ice cooler. Beer? As often as not you will have a bottle plonked down in front of you with no sign of a glass. Too many are content to take their beer in this

(*continued*)

BOX 3.1 CONTINUED

vulgar way. To think of the hours that I and countless others through the decades have spent in researching bubbles, haze, and color!

If only we Stateside (and increasingly those elsewhere who champion the poseur approach of drinking direct from the priciest bottle of premium lager they can find) aspired to emulate the Belgians. I well recall sitting in the home of a friend in Antwerp and, having nominated the beer I wanted from the vast collection offered, waiting for my host to hunt out the correct glass for that specific beverage. That is sophistication. That is the attention to detail that makes Belgians our inspiration for confronting the wine folk head-on.

Recently my student Christine Wright has been exploring folk's perception of wine and beer here in California. She set up stall at a winery in Napa and in a brewery that is probably the most beautiful in the world. Christine asked questions about the relative healthfulness of various beverages—wine and beer, alongside tea, coffee, and soda. In both locations, those polled ranked wine significantly higher than beer. But then Christine pointed out to those questioned that beer contains B vitamins, whereas wine doesn't. That both contain polyphenol antioxidants. That the calorific content of beer and wine is similar. And that the key component in countering atherosclerosis is the alcohol, and it doesn't matter one jot whether the ethanol is from wine or beer. Sure enough, the opinions on beer improved.

Compared to winemakers, brewers have been reluctant to overtly champion their product on a health basis. Certainly it is hard to do that while at the same time boosting the fun elements of the drink. But surely here is a rich terrain for many styles of beer—an opportunity to rein in those people who presently drink no beer but who are prepared to embrace it because it appeals to a high quality of life.

And, for starters, might not brewers point out that wine contains the preservative sulfur dioxide, whereas most beers contain very little?

Then there is the whole myth about wine and food. Some have preached the case for beer as the perfect accompaniment for food. Much as I savor the refreshment of my long slurp of lager as I devour the masala papadums and raw onion, however, I wonder whether it is all a fairy story. My UC Davis colleague and sensory supremo Hildegarde Heymann was quoted in the *Sacramento Bee* as saying "people get so hung up about having the right wine and the right food at dinner that they lose track of the fact that this is a beverage you should enjoy. Drink what you like and eat what you like." In all of her excellent work she has yet to identify a genuine food-drink match. Indeed, for the classic cheese-wine pairing, she found that they mutually interfere rather than reinforce one another.

And so we are left with just that: drink what you like. And dare I suggest that there is a vastly greater span of beers than wines? There is a far broader range of alcohol contents in beers than in wines. Full-flavored beers, subtlest-of-bouquet beers. Diet beers. Every shade of maltiness and hoppiness you can imagine. Bitterness without hop aroma and vice versa. Fizzy beers, flat beers. Beers with fruit and all manner of spices. Wheat beers, rye beers. Draft beers, premium packaged beers. The list goes on.

Wines, by contrast, are grouped as to country, region, varietal, château. For the most part, however, they are subtle variants on a limited theme, generally not straying terribly far from 10%–15% ABV. Reds, wines, and rosé/blush.

Enologists champion vintage, making seasonal inconsistencies into a discussion point, a sales gimmick, an investment. There is no earthly reason why some beers could not be so marketed.

finings. Yeast left over from the fermentation process consumes the sugar to produce carbon dioxide ("natural conditioning"); meanwhile, the finings facilitate the settling out of yeast and other solids. The casks are allowed to rest in the pub cellar before being linked to a hand-pump dispense system. Beer preference is a highly personal matter, but if the author may be permitted to express his view, it is that cask ales are some of the most drinkable beers to be had. A whole movement sprang up in the '70s in the British Isles (the Campaign for Real Ale) to advocate for this product. The reality is that they are perhaps the most demanding of beers to look after (see box 3.2).

INDIA PALE ALE

India Pale Ales (I.P.A.) were designed to have a long shelf life, allowing for their retail in that particular corner of the British Empire. This meant high alcohol, low levels of residual sugars, and plenty of antiseptic hop bitterness. (The Raj recognized the merit of bitter substances to counter diarrhea—hence gin and tonic, with its combination of alcohol and quinine.) They were made with a prolonged boil, to concentrate the sugars so as to allow for a high alcohol content. The first I.P.A. was brewed by George Hodgson in 1822, close to St. Mary le Bow in east London. Soon, Allsopp and Bass launched into the I.P.A. trade.

It is often not realized that, while some bottled beer was shipped, by far the bulk of the beer sold to India was in casks, for bottling locally. Hop bitter acids by no means kill all organisms, and the most prolific inhabitant of those casks bouncing on the ocean waves was *Brettanomyces* (a genus of yeast). The typical flavor notes produced by this organism are "barnyard" or "mousy."

BOX 3.2 THATCHER AND THE DISMANTLING
OF THE BRITISH BREWING INDUSTRY

Margaret Thatcher did many worthy things in her tenure as the British prime minister. But she did nothing for brewing; rather, she presided over the start of the end for the nation's biggest Brewers.

Thatcher was obsessed with choice and avoidance of monopolies. She perceived the vertically integrated British brewing industry as being exactly that. There were six major brewing companies ("The Big Six") and many smaller ones. They all owned their own pubs and sold their beer in their pubs. Thatcher hated that. So she introduced the ruling in the late '80s that a company that brewed beer could own no more than 2,000 pubs (at the time, Bass, the nation's biggest brewer, owned more than 6,000). If, however, you did not brew beer, you could own as many pubs as you wanted. The profit, of course, is made over the bar, unless you are brewing beer in very large quantities, to sell to a distributor/retailer (as happens in the United States). As a result, five of the big six sold out rapidly—Bass, for instance, became Intercontinental Hotels, with more bedrooms (e.g., Holiday Inns) than any other company worldwide. The sixth, Scottish and Newcastle, chose the get-bigger-as-a-brewer route, but even they sold out in 2008.

So now in the United Kingdom there are plenty of pubs to go to (although they are dwindling in number), but many stock the same beers. And traditional products like cask ale are rarer to find, especially good samples, because no longer is there the intimate quality control that was exerted by companies like Bass that had armies of technicians going into the pub cellars and bars making sure that the beer was clarifying and conditioning properly and the pipes were clean.

MILD

This style of English ale is in decline, largely being perceived as an old man's drink! Milds tend to be sweeter and usually (but not always) darker than pale ales. The extra color may sometimes be due to caramel, but is more usually ascribed to a proportion of darker malt. These will be malts that are heated to an extent that affords plenty of color but flavors that are toffee and caramel-like, not burnt and harsh, as milds tend to have mellow flavor. Milds tend to have a lower alcohol content than pale ales (less than 3.5% ABV), and when bottled may be referred to as brown ales.

SCOTCH ALES

These tend to be sweeter, maltier, darker, and less hoppy than English ales. Historically, their strength has been signified in terms of shillings (/-), the old British currency. The stronger the beer, the higher the tax payment to the exchequer in terms of shillings paid. Thus beers could be ranked as low as 60/- and then at 10/- increments through to 90/-. Other traditional terms in Scotland have been *heavy* for their stronger bitters, and *light* for the milds.

BARLEY WINES

These are very strong ales (up to approximately 10% ABV) that are usually sold in smaller volumes, in bottles called "nips." They are made using the strongest "first runnings" from the mash, which may be additionally concentrated by prolonged boiling. Accordingly, not only are the worts that the yeast ferments very strong but they also tend to be richly colored. During fermentation, apart from the high alcohol yield from the plethora of sugars, the high strength of the wort makes for a disproportionate production by yeast of fruit-flavored molecules called esters, the same types of substance that impart the fruitiness to wines.

NATURALLY CONDITIONED BOTTLED BEERS

A relatively few brewers bring their bottled beers into specification for carbon dioxide through "natural conditioning." This is analogous to what happens in the production of cask ales, although there is no addition of whole hops or finings to the bottle. Perhaps the best example is the Sierra Nevada Brewing Company, which carbonates all its beer this way. One of the most famous beers produced through natural conditioning is Worthington White Shield.

PORTER

Porter was first brewed in 1722, a beer born of the industrial age and the first great style produced by the first mega-Brewers in the Industrial Revolution. Taxation in the early-eighteenth-century British brewing industry was on the basis of raw materials rather than beer per se. Darker brown malts kilned over untaxed wood were cheaper than the more heavily taxed paler malts that were cured on coal (the government was concerned about the unhealthy impact of burning coal in urban surroundings). Thus beer brewed from the paler malts fetched twice the price—tuppence—and the London brewers met the budget of the industrial worker (such as the *porters* in the markets of the metropolis) by producing large quantities of beer brewed from the coarser brown malts. It was also argued that London water was too alkaline (because of bicarbonate) to make decent pale ales, extractability of materials (including the less desirable components) from the malt and hops being greater at the higher pH. Then as now, an exceedingly dark, bitter, and richly flavored product could hide a multitude of sins. Plenty of the hops were traditionally used, making porter relatively resistant to spoilage.

STOUT

The advent of Combrune's saccharometer in 1762 meant that brewers could now see the much greater yields of extract that could be derived from paler malts. Hence there was a switch from brown malt to a combination of pale and the black malts that were now being produced in Daniel Wheeler's roasting cylinder (patented 1817—hence the synonymous term for black malt of "patent malt"). The net color of the beer changed from brown to black, with the black malt imparting the burnt, coffee-like characteristics. And the "extra" dark (and stronger) variants were of course "extra stout porters," soon to be simply called *stout*. The modern-day stout remains the only beer (in my opinion) that has its quality enhanced by the use of nitrogen gas. The nitrogen shaves the harshness that is delivered by roasted malts and roasted barley, while at the same time yielding a quality of foam that appears entirely natural on a black beer but not on a pale ale.

IMPERIAL STOUT

Also known as "Russian Imperial Stout," it is a strong (9%–10% ABV) and therefore warming dark beer originally brewed for export to the court of the Tsar of Russia.

SWEET STOUT AND OTHER STOUTS

Milk (or sweet) stout was so named for the whey-derived lactose employed in a production process dating back to 1669. Lactose is not metabolized by brewer's yeast, and it has only one-fifth of the sweetness of sucrose. Apart from this modest contribution to sweetness, lactose also affords body. There are oatmeal stouts, with perhaps 5% of the grist comprising rolled oats, which are somewhat dry to the palate. Stout has long been associated with oysters. Primarily this was in the form of stout accompanying a plate of the shellfish, but from time to time there have been oyster stouts, with the beer allowed to "rest" on a bed of oyster shells or, in later years, suffused with an oyster flavoring. Stout plus bitter ale (traditionally Guinness and Bass) yields *black and tan*, while *black velvet* is stout and champagne.

GERMAN ALES

Alt means "old," and *altbier* (often abbreviated alt) is "brewed in the old way," from a grist of barley and (to a lesser extent) wheat malts. Alts tend to be a dark copper color and relatively bitter.

Kölsch, from Cologne, is a much lighter beer, with the darker malts used in the production of alt replaced by the less intensely kilned Vienna malt.

The German wheat beers are made with top-fermenting yeast. *Weizenbier* is made from a grist comprising at least 50% wheat malt. It tends to be relatively highly carbonated, with pronounced fruity notes but also a distinct clove-like character that derives from the special yeasts used to make such beers. Absent such a clove-like note, a weizenbier is not authentic. Such beers tend to be relatively pale or straw-like in appearance—and, to be authentic, should strictly *never* be served with a slice of lemon. Such beers are most frequently *hefe-weizens* ("yeast-wheats") and are cloudy due to the presence of a yeast residue, which is traditionally employed to carbonate the bottled product through natural conditioning. The term *kristallweizen* denotes that the beers have been filtered to remove yeast.

Weissbier ("white beer") is much weaker (e.g., 2.8% ABV) and is made from a grist of less than 50% wheat malt. Lactic acid bacteria are used to generate a low pH of 3.2–3.4 and therefore much sourness. These beers tend to be taken with a dash of raspberry or sweet woodruff syrups.

BELGIAN TOP FERMENTATION BEERS

There is a greater diversity of beer styles in Belgium than in any other country.

One of the famous genres is *Trappist beers.* These are relatively dark, intensely bitter, and acidic, with clear fruity notes and strengths as high as 12.5% ABV.

A very distinctive Belgian beer style is *lambic,* a variety of which is *gueuze.* These products have complex flavor characteristics due to the metabolic activities of diverse microflora beyond brewing yeast alone. They tend to be quite sour (low pH) and are frequently not clear. It is customary to enhance the drinkability of such beers by adding various flavorants such as cherries (*Kriek*), raspberries (*Framboise*), blackcurrants (*Cassis*), and peaches (*Peche*).

Figure 3.1 gives a generalized overview of some of the top fermentation beers on scales of maltiness and hoppiness.

BOTTOM FERMENTATION BEERS

PILSNER

What is perhaps the classic style originated in mid-nineteenth-century Bohemia, in the Pilsen city (Plzeň) brewhouse. It is quite a malty brew, typically with 4.8%–5.1% ABV and a pale gold color. Particularly prized is the late hop character.

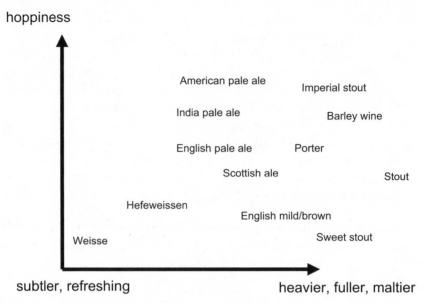

FIGURE 3.1 A rating of ales by maltiness and hoppiness

All too often the term "lager" is used synonymously with "pilsner" (or "pils"). Lager as a term is really an umbrella description for all relatively pale brews, fermented and dispensed at low temperatures. However, it should not be forgotten that some lagers are very dark, as we shall see. It is difficult to discern whether we truly have a lager or an ale. For instance, there are some brewers who use only a single strain of yeast to ferment all their brands, whether ales or lagers.

BOCK

As a style, bocks tend to be stronger than pilsners, at 6%–8% ABV. They typically have sulfury and malty flavors with colors that range from straw to dark brown. Doppelbocks may contain up to 12% ABV.

MÄRZEN

"March beer" is brewed in March for consumption in the ensuing September (*Oktoberfest*). The style originated following the edict that brewers were forbidden to brew in the summer months. Accordingly, the brews were historically of relatively high strength (up to 6.5% ABV). There are pale and dark versions.

HELLES

The German word *hell* means pale. These pale amber lagers tend to be very malty, of relatively low bitterness and hop character, and around 4.5%–5.5% ABV.

DUNKEL

These have comparable flavors and strengths to helles, but are copper-brown in color.

SCHWARZBIER

As the name indicates, these are black lagers. They are dry, with alcohol contents between 3.8% and 5% by volume.

RAUCHBIER

These are produced from a grist that incorporates smoked malts and thus they have aroma notes comparable to those of the peated malt whiskies.

MALT LIQUOR

Malt liquors in the United States are products of relatively high alcohol content (6%–7.5% ABV) that are very pale, very lightly hopped, and quite malty and sweet. For the most part, it is a label rather than a style of lager.

STEAM BEER

The origins of this product can be traced to the California gold rush and a demand for light and refreshing drinks despite the unavailability of ice for cold storage and conditioning. Bottom yeasts were used at warmer fermentation temperatures than were customary for lagers, in shallow vessels into which the "steaming" wort was introduced to cool.

Figure 3.2 gives a generalized overview of some of the bottom fermentation beers on scales of maltiness and hoppiness.

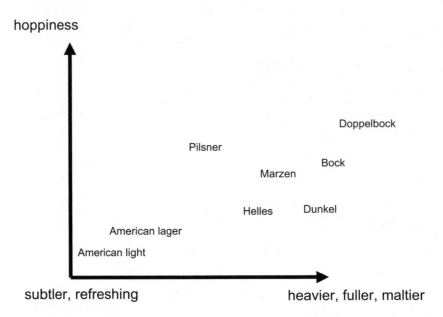

FIGURE 3.2 A rating of lagers by maltiness and hoppiness

OTHER BEERS

I frequently make the point in championing the worth of beer over wine that there is far more diversity in the former. So, while the wine world (with the exception of fortified stuff) restricts itself to red, white, and pink, for beers we can have everything from colorless to black and every flavor imaginable.

A dear friend of mine firmly reminds me that it is a case of each to his or her own, and what she enjoys is her business. Although I would never say it to my buddies in the wine world, my personal preference would be for brewers to take a leaf from their book and champion the diversity that different malts and hops can afford to beers within a relatively few styles of ales and lagers. Instead there seems to be no limit to the ingenuity (foolhardiness?) of brewers as they expand their portfolios with ever wackier inventions.

Having said which, some of the more extreme products purchased across a bar have been with us for a very long time. From my youth in England I well recall *shandy,* in which pale ale and lemonade (more 7 Up than mom's own freshly squeezed stuff) were mixed in equal quantities. Refreshing after playing a sport. Some folks used to add a dash of blackcurrant juice to their lagers, or a splash of lemonade ("lager top").

A well-known English brewer developed a famous *chocolate stout,* in which substantial chocolate malt was used, but also a bar of chocolate and some chocolate-flavored essence. Coriander, chilies, pumpkins…there is no limit to what brewers have used. I recall a recent sampling of a beer containing ginseng, caffeine, and guarana. As my irises dilated, the chap who had offered it to me smirked and said "it wasn't designed for you, Charlie." This was a party beer.

Apart from shandy in the United Kingdom, and the comparable *Radler* in Germany and *panaché* in France, there are several other traditions in the world of mixing beer with other things. One of the best known is *Michelada* in Mexico, in which lager is mixed with tomato juice and salt and often Worcestershire sauce or even soy sauce or teriyaki sauce.

A genre on the cusp between beer and wine is the *malternatives,* sometimes known as flavored alcoholic beverages. They are basically unhopped beers, with the color removed and a diversity of flavors introduced. They are produced in this way for tax purposes—they are beer-based and with a strength typical of beers, and as such are taxed in many states as beers, which is at a lower rate than spirits. There is a strong lobby to levy tax on them as spirits, for it is believed in many quarters that these relatively sweet products are targeted at the younger element, those who have not yet acquired a taste for bitter products.

ICE BEERS

The ice beer story is a fascinating example of how an entirely new beer concept emerged from a technology that failed to serve the purpose for which it was originally installed. In the 1980s, many Brewers had decided that, rather than ship finished beer around the countryside to its destination, it would make economic sense to transport the beer in a concentrated form and then reconstitute it at the point of sale. They experimented with a technique called "freeze concentration" that took advantage of the fact that if you freeze beer, the first thing to come out of solution is almost pure water, namely ice. Most of the beer components remain in solution in a concentrated form.

Labatt, a major Canadian Brewer, was one company that experimented with the technique. They quickly realized that it wasn't going to be a winner for the purpose for which it was intended. Fortunately for Graham Stewart, their technical director at the time, and his colleagues, they hit upon an even more exciting use for freeze concentration. They were looking for a new angle on beer marketing and identified *ice* as being a powerful concept that associated extremely well with beer in the perception of Canadian drinkers. It didn't take long for the intellectual leap to be made: "Hey, let's chill out our beer and position a new beer genre as 'ice beer.'" As Professor Stewart says: "After all, Canadians already knew all about putting beer out onto the window ledge in the winter, freezing ice out from it, thereby increasing the alcohol content!"

By the early 1990s a new and exciting beer story was being told, and most major Brewers developed their own ice brands. In 1996 some 24 million barrels of ice beer were brewed in the United States, with the market share for such beers increasing by almost 4% on the previous year.

DRY BEERS

The mid-'80s saw the emergence of dry beer, and through it the astonishing growth of the Japanese Brewer Asahi. They launched a new brand called Super Dry and saw a 25% increase in their market share within three years. As the name suggests, it is a straightforward concept analogous to dry wine—a lager with a relatively low proportion of residual sugar—but clever marketing, and the characteristic outstanding package quality associated with Japanese Brewers, made it a clear winner. It was a product deliberately designed to appeal to as many people as possible through having no extreme flavor characteristics that might alienate sections of the populace. In no time it was followed by other dry beers ("me too's"), and a dozen countries contributed over 30 new brands of dry beer.

LIGHT BEERS

Premium light beers now constitute the most popular beer category in the United States and have come a long way from the first reduced-calorie brew by Rheingold, called Gablinger's. These beer styles are differentiated by their content of residual carbohydrate: standard premium beers contain a proportion of carbohydrate that survives the fermentation process, whereas a light beer has most or all of this sugar removed by techniques we will visit in chapter 9. Therefore, these beers have lower calorie contents, provided they don't contain extra alcohol, which in itself is a contributor to calorie intake. Thus it is perhaps no surprise that in a market (the United States) where 24% of all beer is consumed by women, the proportion of light beers drunk by women increases to 30%.

DRAFT BEERS

The word *draft* (*draught* in the British Isles) can refer to two entirely distinct beer types. Traditionally, it refers to beer dispensed from kegs or casks via pipes and pumps, or indeed straight from the cask, as is still the case for some of the traditional English ales. It is also used, however, to describe canned beer that has not been pasteurized but rather sterile-filtered. The marketers had a new angle for canned beer: "as nature intended." Much beer worldwide is now marketed on this angle of "non-heat treated." For instance, no beer in Japan is pasteurized. The fact is that, provided the oxygen levels in the beer are low prior to pasteurization, this process has no adverse impact on flavor and is actually to the benefit of foam, which can deteriorate with time in non-pasteurized beers. It certainly is a curiosity that the lack of pasteurization is taken by some as a positive thing. I can't imagine customers caring much for this if we were talking milk!

FROM CASK-CONDITIONED TO NITROGENATED BEERS

The big growth market for beers in the United Kingdom is in nitrogenated products. Their emergence is an informative lesson in how modern technology can throw up products whose origins are in traditional practice.

The classic beer style in England is nonpasteurized ale of relatively low carbon dioxide content. Happily, many famous brews of this type continue to be produced. The production of traditional English ales involves them going from fermentation into casks, to which are added hops, sugar, and fining materials that help the residual yeast to settle out. That yeast uses the sugar to carry out a secondary fermentation, which carbonates the beer to a modest extent. The product is not pasteurized, and must be consumed within a few weeks. It is characterized by a robust hoppy flavor but also by much less gas "fizz" than other product types.

Again in the mid-'80s, and with the projected demographic shift to more drinking at home rather than in pubs and bars, marketers in the United Kingdom decided that they would really like to be able to sell this type of beer in cans for domestic consumption. The problem had to do with the low CO_2 content, for the gas is generally required to pressurize and provide rigidity to cans, and also to put a head on beer. For cask beers, it is the hand pump characteristic of the English pub that does the work in frothing the beer. For "normal" canned beers, the relatively high gas content does the job for you on pouring. So how could the foaming problem be overcome for canned beers containing relatively little carbon dioxide? The answer was the "widget," a piece of plastic put into the can that flexes when the can is opened and causes bubbles to come out of solution (see chapter 4). This technology had been invented by Guinness, a Brewer with a long tradition of producing stouts with superbly stable heads. Allied to this was the realization that nitrogen gas makes vastly more stable foams than does carbon dioxide, again a technology that had been pioneered by Guinness and taken advantage of by many Brewers to enhance the heads on their draft beers. So nitrogen was included in the cans—dropped in at canning in its liquid form. Not only does the nitrogen help the foam, but it also smoothes out the palate, enhancing the drinkability of some of the beers that contain it, notably the stouts. The sales of canned beer with widgets zoomed, and, seeing this, Brewers recognized the potential for so-called nitrokeg beers, where the beer is on draft dispense but is characterized by low CO_2 and the presence of N_2. The merits of nitrogen and widgets were not appreciated by every brewer. Tired of complaints about the canned beer having taken a turn for the worse since the introduction of the "lump of plastic," one English company reversed matters, eliminated the device, and proudly announced on the label "widget-free ale." Applause from this author, for one.

NON- AND LOW-ALCOHOL BEERS

"Normal" beers range in their alcohol content from 2.5% to 13%. To a Bavarian used to beers having 6% alcohol or more, the regular tipple of the English ale drinker at, say, 4% might be viewed as "low-alcohol." Non- and low-alcohol beers (NAB/LABs) can be classified in many ways. For our purposes, I will define them as beers containing less than 0.05% and less than 2% alcohol (by volume), respectively.

While there are a few successful NAB/LABs in the world, they are the exception rather than the rule. For many people it is a contradiction in terms to associate a beer with low alcohol: after all, what is a beer if it doesn't deliver a "kick"? The rationale behind developing such beers in the first place is an interesting one, and is largely based on the proposal that peer pressure amongst drinkers convinces some people of the need to be seen to be drinking a product indistinguishable (by sight) from a normal beer, but one which is of reduced alcohol content, thereby enabling them to drive. Increasingly, it has been appreciated that this peer pressure phenomenon was overstated and that educated consumers will happily drink an established nonalcoholic product, say a juice or a cola, if the circumstances demand it. It seems that the only justification for purchasing a beer of low alcohol content is if it is pleasing to the palate. And that certainly hasn't always been the case for many such beers. The shortage of quality products in this genre is reflected in the statistics: in the United Kingdom NAB/LABs grew to occupy 1.1% of beer sales in 1989, but this had declined to 0.3% of sales just six years later.

This type of product has been made in many ways. Perhaps the most common techniques are limiting alcohol formation in fermentation and stripping out the alcohol from a "normal" beer. In the first case, the yeast can be removed from the fermenting mixture early on, or indeed the wort that the yeast is furnished with may be produced such that its sugars are much less fermentable. Alcohol can be removed by reverse osmosis or by evaporating off the alcohol using vacuum distillation. It should come as no surprise that attempts to remove alcohol will also result in the stripping away of desirable flavors. Equally, if fermentation is not allowed to proceed to completion, these very flavor compounds are not properly developed and undesirable components derived from malt are not removed. Either way, the flavor will be a problem. And considering that ethanol itself influences the flavor delivery of other components of beer, as well as itself contributing to flavor, it will be realized why good NAB/LABs are few and far between.

THE OTHER EXTREME

The world record for strong beer is currently with Utopias, from the Boston Beer Company, weighing in at 25.5% ABV. Aged in wood and packaged into 700-mL bottles shaped like brewery kettles, it retails (as I write) at $130 a go. They say that "it should be savored like an old sherry, vintage port, or fine cognac." I guess so, at that price. I ask myself "Why?" I guess the retort would be, like George Mallory tilting at Everest, "Because it is there."

What diversity! A rich seam of products, yet all of them definable by a range of quality criteria that speak to the appearance and flavor of the product. Let's take a look at such things.

EYES, NOSE, AND THROAT

THE QUALITY OF BEER

The consumer of beer drinks as much with her eyes as with her mouth. Certainly beer drinking can be as much a visual pleasure as it is thirst-quenching. The quality attributes of beer perceived by the eye can influence our perception of flavor, as demonstrated by a simple experiment. Try adding a few drops of a flavorless dye (the type you may use in your kitchen) to a lager so that the color darkens to that more typical of ale. People presented with this beer will judge its flavor to be closer to that of an ale than a lager, whereas if they are blindfolded they certainly won't be able to tell apart the taste of the beer before and after the dye has been added.

Color is just one visual quality parameter of beer. Most people (other than those smitten with hefeweizens) prefer their beers to be sparkling bright, with no suggestion of cloud or haze. However, there is greater variation in the extent to which drinkers like a head of foam on beer. In some countries a copious delivery of foam on dispense is essential: for instance, it is traditional in countries such as Belgium for as much as half the contents of the glass to comprise froth. In the United Kingdom, there are distinct regional differences: in some places, for example London and the South East, foam frequently seems to be regarded as an inconvenience. By contrast, a stable head, perhaps two inches deep, is generally required in the north of England. But unlike, say, the Belgian, many an Englishman appears to want foam *and* a full measure of beer. Matters reached a head (one might say) when the status of beer foam was challenged in the courts of law. Those insisting on a full pint of liquid challenged landlords who dutifully dispensed the beer with a head. The High Court judged that a reasonably sized head *should* be regarded as an integral feature of the beer, but that the customer is within his rights to insist on a full measure of liquid beer.

Such concerns are only relevant, of course, for draft beer. For beer in cans and bottles the unit of volume is fixed by what is present in the container. Whether a head is generated or not is more often in the hands of the customer rather than the bartender. Indeed, that's if the beer gets poured at all. Some prefer their booze directly out of the bottle or can, in which case foam (and color and clarity, for that matter) assume a more academic dimension.

In the same way that color influences the perception of the flavor of a beer, so too does the head seem to affect a drinker's judgment. Again, there is undoubtedly a psychological component at work here. It is, however, likely that the presence of a foam does have a direct bearing on the release of flavor components from the beer; in other words, a beer will smell differently when it displays a head of foam as opposed to when it does not. Not only that, but there are also substances present in beer that have a tendency to move into surfaces such as the bubble walls in foam and are therefore called "surface-active compounds." These include the bitter compounds, and so the foam has proportionately more bitterness than has the rest of the beer.

You can see, then, that long before a drinker raises the glass to their lips, they will have already made some telling judgments on its quality, drawn from visual stimuli alone: the quality of the can or bottle, the "font" if the beer is on draft dispense, the appearance of the foam, the color, and whether the beer is cloudy. And all this is quite apart from the effect of other stimuli associated with the place in which the beer is being drunk: the lighting, the background music, the attractiveness of the bar layout, the foodstuffs being consumed alongside the beer, and even the company being kept! Beer flavor is important, of course, but even the most delicious of beers won't be enjoyed if all the other elements of the drinking experience are flat.

FOAM

Typically a packaged beer contains between 2.2 and 2.8 volumes of carbon dioxide (i.e., for every milliliter of beer there are between 2.2 and 2.8 milliliters of CO_2 dissolved in it). At atmospheric pressure and 0°C, a beer will dissolve no more than its own volume of CO_2. Introduction of these high levels of CO_2 demands the pressurizing of beer. Yet if you take the cap off a bottle of beer, the gas normally stays in solution. The beer is said to be supersaturated. To produce foam you must do some work.

Foaming is dependent upon the phenomenon of nucleation, that is, the creation of bubbles. Bubble growth and release occurs at nucleation sites, which might include cracks in the surface of a glass, insoluble particles in beer, or gas pockets introduced during dispense. Pockets of gas are introduced whenever beer is agitated, as anyone who has tried to open a dropped can of lager will tell you.

The physics of bubble formation is far from completely understood, and is also astonishingly complicated. Brewers have approached the problem as much empirically as on a firm scientific foundation. For instance, glasses have been scratched to ensure a plentiful and continuous release of gas bubbles to replenish the foam, a phenomenon sometimes referred to as "beading." Draft dispense is typically through a tap designed to promote gas release. Most recent of all has come the so-called widget. Widgets have even found their way into bottles.

Although beers are generally supersaturated with CO_2, foam generation is still easier and more extensive the more highly carbonated is the beer. The formation of foam is encouraged by ethanol, though the more alcohol in the beer (all other things being equal), the worse the foam stability.

Various physical factors are involved in dictating the rate at which beer foam collapses. As soon as foam has formed, beer trapped between the bubbles starts to drain from it because of gravity. Anything that increases the viscosity of the beer should reduce the rate of drainage. Since viscosity increases as temperature decreases, colder beer has better foam stability. Counter to this is the fact that foam *forms* more readily at higher temperatures, because gas is less soluble.

As liquid drains, the regions between bubbles become thinner, leading to coalescence as bubbles merge into bigger ones. The effect is to coarsen the foam and make it less attractive: foams with smaller bubbles are whiter, with a more luscious consistency in the mouth.

The least desirable set of circumstances occurs if the bubbles in foam are of assorted sizes. The gas pressure in a small bubble is greater than that in a larger one. If two such bubbles are next to one another, gas will pass from the small bubble to the larger one until the smaller bubble disappears. The result, once again, is a shift to "bladdery" and unattractive foam, as well as one with half as many bubbles. This phenomenon, which is called "disproportionation," happens more quickly at higher temperatures, but to a lesser extent if the gas pressure above the liquid is increased. Try covering your beer glass: you'll find that the foam survives longer. This is the principle of the German beer stein, although as steins are generally ornate and the beer can't be seen, the objective is rather defeated!

The rate of disproportionation is also less for gases of lower solubility. For instance, nitrogen is only sparingly soluble in water. Inclusion of just 20–50 mg of nitrogen gas per liter of beer leads to foam with very small bubbles, a foam that is therefore extremely creamy and stable.

Of course, when bubbles are formed in a liquid the effect is to increase the surface area. This opposes the forces of surface tension (the force that makes a liquid such as water occupy the lowest possible surface area in proportion to volume— hence droplets), and it is for this reason that pure liquids can't give stable foams. Materials must be present that are able to get into the bubble wall to stabilize it.

In beer, the backbone material for bubbles is protein, which comes from the malt. In particular, it is those proteins that have a relatively high degree of hydrophobicity (water-hating character) that preferentially migrate into the head. There they encounter other substances with high hydrophobic character, notably the molecules from hops that give beer its bitterness (see below). The interactions between the proteins and the bitter substances hold the bubbles together. This interaction is not spontaneous, and it proceeds over a period of minutes. As it happens, the texture of the foam changes from liquid to almost solid, in which state foam can adhere to the glass surface, a phenomenon known as "lacing" or "cling." The longer you delay slurping your beer, the greater the opportunity for the textural transition to occur, and therefore the better the lacing.

Just as there are materials in beer that promote foam, so there are other substances that interfere with it by getting in between the protein molecules and preventing them from interacting. These materials include ethanol (see above) but are primarily lipids (including fats), which, like the proteins, can originate from the malt. However, good brewing practice should ensure that very low levels of lipids survive into the beer. It is much more likely that these types of substance will get into the beer when it is in the glass and destroy the foam. Any grease or fats associated with food are bad news for beer: if you eat potato chips, the oils associated with them easily kill foam. Lipstick, too, contains waxy substances that will pop bubbles, and the detergents and rinse-aids used to wash glasses also tend to be foam-negative. When beer glasses are washed, the detergent must always be washed from the glasses using clean water and the glasses preferably allowed to dry by draining. If the glasses are wiped on a kitchen cloth, it must be a clean one.

BOX 4.1 FOAM STABILIZER

Alginates are polysaccharides extracted by alkali from brown seaweed, harvested on the coasts of North America, Scotland, Ireland, Norway, France, Japan, China, Korea, Chile, South Africa, and Australia. Of these, China easily generates the most, though the United Kingdom and United States are the biggest processors, with one company dominating at 70% of the market.

Insoluble alginate from the raw plant is solubilized and filtered and the alginic acid obtained is esterified by reaction with gaseous propylene oxide. After washing with alcohol, the product is dried and milled.

Alginates, esterified or not, are very widely used in industry. About 50% goes to textile printing, with 30% headed to food use. Ice cream, sherbets, milk shakes, yogurts, icings, cake and pie fillings, meringues, glazes, salad dressings, and noncarbonated fruit drinks all contain the stuff. Beers incorporating PGA, then, are in pretty respectable and wholesome company.

Before we leave foam we should remember that it isn't always good news. From time to time foaming occurs spontaneously when a can or bottle is opened. In extreme examples, as much as two-thirds of the contents spew forth in a wild and uncontrollable manner. Most people find this to be somewhat irritating! There may be several reasons for the phenomenon, which is called *gushing*. The first, of course, is that the package has been ill-treated, dropped, or shaken. Brewers take great care when shipping beer to avoid unnecessary agitation of the beer. And provided a beer is given an hour or two to settle after being dropped or shaken, then the beer won't be wild when the can or bottle is opened.

Unfortunately, gushing is sometimes due to substances that promote the phenomenon and that originated in the raw materials. Barley grown in wetter climates is susceptible to infection by a fungus called fusarium. This produces a very small protein molecule that gets into malt and from there into beer, where it acts as a very active nucleation site for bubble formation. Another type of molecule that can act in the same way is an oxidation product of hops, which is from time to time to be found in certain preparations used to bitter beer.

COLOR

An enormous range of colors is found in beers: there are exceedingly pale straw-like lagers, copper-colored ales, rich brown milds, and the blackest of stouts. Some years ago a North American brewing company even made a beer that looked like water in its clear glass bottle. It wasn't on the market for very long.

This wonderful range of colors is seldom achieved by the addition of coloring materials, although caramels have been and continue to be used in some quarters for this purpose. Generally the malt and other solid grist materials that are used in the brewhouse determine the color of beer. However, recently a new method of coloring beers has been introduced in which the color of dark malts is extracted and separated from the flavor-active molecules in those malts for addition as a liquid late in the brewing process. This extraction process involves making an extract of the dark malt in water and fractionating it according to the size of the substances it contains. This can be achieved using special membranes that allow small molecules to pass through but big molecules to be retained. The components responsible for flavor are small, but the coloring materials are large. Using this technique, then, preparations have become available that enable a beer to be made darker without introducing the smoky/burnt characteristics typical of a roast malt but also, conversely, to introduce such flavors into pale beers without making them dark. This presents splendid new product development opportunities to Brewers, as well as presenting an opportunity for introducing color without the use of caramel.

The color-forming materials in the grist are primarily complex molecules produced when sugars and amino acids are heated. The more intense the heating regime, the darker the color produced. Heating is an integral feature of the process by which malt is produced (chapter 5). The more intense is this kilning, the darker will be the malt. Also, the more sugars and amino acids are present, the greater is the potential for making these colorful molecules. Sugars are formed during the germination of barley when complex carbohydrates are broken down. Similarly, the amino acids are the end point of protein breakdown. A malt destined for lager production tends to have had relatively limited germination and, more significantly, is kilned to modest temperatures, and so the color contribution from it is low. Ale malts are more extensively modified during germination and are kilned to a higher temperature, so they are darker. If the malt is kilned to ever more intense extremes, then profoundly dark malts are obtained. Such materials are traditionally employed in the production of darker ales and stouts.

A second source of color in brewing is the oxidation of polyphenol or tannin materials. These tannin-type molecules originate from both malt and hops and are prone to oxidation if large amounts of oxygen are allowed to enter into the brewhouse operations. The reaction involved is exactly analogous to the browning of sliced apples. If this source of color is to be eliminated, it is essential that oxygen must be excluded in the mash mixer and, especially, the wort kettle (see chapter 8).

HAZE

Oxidation of polyphenols is much more important for another reason: it results in the formation of haze. In the oxidation process, the individual tannin units associate to form larger molecules that associate with protein to form insoluble particles that cause turbidity in beer. The reactions involved are similar to those responsible for the tanning of leather.

Other materials may cause cloudiness in beer. For instance, if the complex carbohydrates of barley, chiefly the starch, or the polysaccharides that make up the barley cell walls are not properly digested in malting and brewing, they can precipitate out of beer as hazes or gels, particularly if beer is chilled excessively, for example in the case of inadvertent freezing. Another natural component of malt is oxalic acid (the same stuff that furs your tongue when you eat rhubarb), which brewers should ensure is removed in the brewhouse operations by having enough calcium in the water to precipitate it out. If they fail in this task, the oxalic acid will survive into beer. This is primarily a problem for draft beer, because oxalate will precipitate out in the dispense lines and clog them. This is so-called beer stone.

FLAVOR

The flavor of beer is no simple affair. There are the obvious tastes one associates with the product, in particular the bitterness imparted by hops. And, as for most foodstuffs, the characters of sweet and salt play a part, although few desirable beers have sourness among their attributes. A wide range of volatile substances contributes to the aroma of beers, including esters, sulfur-containing compounds, and essential oils from hops. Ethanol itself provides a warming effect and seems to influence the extent to which other molecules contribute to a beer's character. Even carbon dioxide has a role to play.

To complicate matters further, it should be appreciated that the flavor of beer is not static. From the time that fermentation is complete to the moment that the beer is packaged, changes occur in its taste and aroma. And it doesn't stop there; just as with wine, the character of beer changes in the package. Only rarely are these changes for the better in beer, as we shall see.

THE NATURE OF BEER FLAVOR

The flavor of a beer can be broken down into its taste, smell (aroma), and mouthfeel (texture or body) (figure 4.1). Brewers also talk about "flavor stability," in reference to the deterioration in quality as the beer ages in package.

TASTE. The bitterness of beer is due to a group of compounds called the iso-α-acids that are derived from precursors in hops (see chapter 7). A drinker's perception of bitterness changes after she has taken a sip of beer. There is an initial surge in perceived bitterness, followed by a gradual subsidence of the effect. In fact, the ability of a drinker to estimate the bitterness of a beer is generally fairly poor. A well-trained taster may be able to address the problem, but everybody else will tend to have their judgment clouded by other features of the beer, most notably sweetness.

The sweetness of beers is due to residual sugars that have not been fermented into alcohol. Frequently the brewer will add sugar ("primings") to the beer before packaging to ensure a desirable sweetness-sourness-bitterness balance.

Although "salty" is not a word that many people would use to describe beer, certain salts do contribute to beer flavor. Sodium and potassium in the beer impart the taste that you and I know as "salty." However, in particular the ratio of chloride to sulfate is felt by some to be important. Sulfate is claimed to increase the dryness of a beer, while chloride is said to mellow the

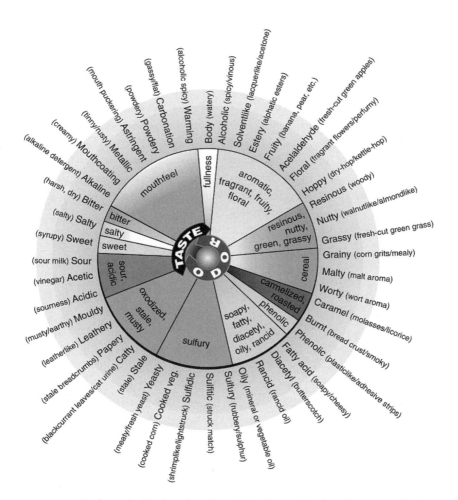

FIGURE 4.1 The flavor wheel. Redrawn from the original with permission from the American Society of Brewing Chemists. The wheel presents the breadth of flavor terms that a trained taster may use to "score" beer aroma and taste.

palate and impart fullness. The importance of this chloride-to-sulfate ratio is one example of the plethora of dogmatic beliefs held by many brewers. In fact, there is little if any published scientific data to justify the conclusions made concerning this ratio—which is not to say that it is not important, but rather that, if it were in the dock in a court of law, it wouldn't have much of a hard-and-fast defense.

Other ions are certainly significant. For example, traces of iron that might be picked up from the materials used to filter beer will give an unacceptable metallic character to a product. ("Metallic" flavor in beer can also be due to some compounds that don't actually contain any metal. A curious test for metallic character in beer is to dip your finger in the beer and rub the liquid on your other hand, around where your thumb meets the fleshy part on the hand. Any metallic character can be smelled.)

AROMA. People often refer to the taste and the smell of a foodstuff, such as beer. In fact, these are very closely related sensory phenomena. What a drinker will generally describe as the "flavor" of a beer is in reality a character that is primarily detected in the nasal passages; it is strictly speaking defined as aroma. If you have suffered from a head cold and have had blocked nasal passages, you will appreciate the effect that this can have in eliminating the sense of taste. Actually, there is an increasing realization that there is more than just sweet, salt, sour, and bitter detected on the tongue.

Many different types of molecule influence aroma. Lots of these are produced by yeast during fermentation. There are alcohols other than ethanol, which can impart coconut or solvent-like characters, and there are aldehydes, which give aromas like green leaves. Principally, though, there are the esters, the short-chain fatty acids, and many of the sulfur compounds. The pH of beer is also largely dependent on fermentation, as yeast acts to lower the pH of wort from 5.2–5.5 to 3.8–4.5.

The pH of beer has enormous influence on product quality. Many of the molecules in beer can exist in charged and uncharged forms, the relative proportions of which are directly dependent on the pH. For instance, the bitter compounds exist in both uncharged and charged states, and the former is bitterer than the latter. The hydrogen ions that cause a beer to be acid (low pH—see Appendix) impart sourness. That is, if a beer has a low pH (say 3.8) it will be sourer than one of pH 4.5.

A selection of esters is present in beers, imparting flavors like pear, banana, guava, pineapple, and roses. It is the mix of esters and other volatile compounds that determines the aroma of beer. Drinkers might say that a beer is "fruity" or even "bananas," but its true character is seldom so simple as to be traceable to just one or even a very few types of flavor molecules. Rather, it is the combined effect of a complex mixture that will determine overall "nose." Such complexity ensures that individual beers are unique, but it also makes considerable demands on the brewster if she is to ensure consistency.

Esters, then, have a range of individual aromas in the pure state, which aren't simply related to the character that these substances impart to the complex matrix that is beer. The same applies to other classes of compound found in beer. For instance, various sulfur compounds may be present, giving aromas such as rotten egg, canned corn, onion, leek, cooked vegetable, garlic, and mashed potato. These don't sound very appealing, but the reality is that these notes are seldom overt but, when melded into the entire portfolio of flavors present in a beer, deliver an overall experience to generally delight the consumer.

Short-chain fatty acids generally provide undesirable characters to beer. Descriptors include cheesy, goaty, body odor, and wet dog! Happily for the drinker, the Brewer's control over the process means that undesirable levels of this type of compound seldom find their way into the beer.

Just as undesirable is the character introduced by the so-called vicinal diketones. Diacetyl, the most important of these, has an intense flavor of popcorn (indeed, until the practice was recently outlawed, copious quantities of diacetyl were used in the manufacture of the favorite nibble of the movie theater. Like that smell or loathe it, few of us like our beer to smell of popcorn!)

Diacetyl is naturally produced in all brewery fermentations. It is an offshoot of the metabolic pathways that the yeast uses to make some of the building blocks that it requires for growth. The diacetyl leaks out of the yeast cell and into the fermenting broth. Happily, yeast is also capable of mopping up the diacetyl again. And so, toward the end of fermentation, the yeast scavenges the diacetyl and converts it into substances that do not have an intense aroma. To do this the yeast must be in a healthy state, but even then the process can take a considerable period of time. Thus the period for which a beer must stay in fermenter depends not only on the time taken to convert sugar into alcohol, but also on the additional days required to eliminate diacetyl and take it below its flavor threshold, which may be exceedingly low in a gently flavored American lager.

Another undesirable flavor in beer is acetaldehyde, which imparts a character of green apples. Surpassing all others in the undesirable flavor stakes, however, is the aroma imparted by a compound called 3-methyl-2-butene-1-thiol (MBT). This is produced through the degradation of the iso-α-acid bitter substances, a breakdown brought on by light. People differ in their sensitivity to MBT, but for many it can be detected at levels as low as 0.4 parts per trillion. To put this into perspective, these poor people would have been able to detect a tenth of a gram of MBT distributed throughout the balloon of the airship Graf Zeppelin II. The aroma that MBT imparts is referred to either as "lightstruck" or, worse still, "skunky." Brewers have known about the problem for many years, and it is the reason beers need to be protected from light. Brown glass is better than green glass in this context. Clear flint glass is the worst option. An alternative strategy is to use modified (reduced) iso-α-acids that, when broken down, no longer give MBT. The added advantage is that this reduction enhances their foam-stabilizing properties, although the texture and appearance of such foams is perceived by many as being artificial.

It's high time we returned to some of the more desirable characters in beer. Generally these originate from the malt and hops. Malty character is quite complex, and is due to a range of chemical species. Hoppy character, too, is far from simple and may, indeed, take various forms. It is due to the essential oils of hops, but the contribution they make to aroma differs considerably between beers.

For the most part, the hops are added at the start of the boiling process. The essential oils, being volatile, are comprehensively driven off during the boiling operation, and the resultant beers, while bitter, have no hoppy character. Because of this volatility, some Brewers hold back a proportion of the hops and add them late in the boil. In this way some of the essential oils survive into the wort and thence into the beer. The process, for obvious reasons, is called "late hopping."

In traditional ale brewing in England, with the beer dispensed from casks, it is customary to add a handful of hop cones to the beer at the point of filling. In this way all of the essential oils of those hops have the opportunity to enter into the beer, affording a complex and characterful nature to such products. This is "dry hopping."

FLAVOR BALANCE. This is an appropriate stage to emphasize that beer flavor is not simply a matter of introducing greater or lesser quantities of a given taste or aroma into a product, depending on the character desired in a given product. A beer is pleasing, interesting, and above all *drinkable* because it has its various organoleptic properties in balance.

If I were to have a single criticism of some of the beers being produced by microbrewers, it would be that they have failed to grasp this point. Many of these Brewers seem to have overdosed on hops, rendering their beers intensely aromatic and, of course, extremely bitter. It is perfectly satisfactory to have a very bitter beer—many such products of long standing exist in the world. Equally, there are many leading brands of lager that have pronounced late hop character, but possess modest levels of bitterness. We have seen above how aroma and bitterness levels in beer can be independently adjusted, enabling products with excellent *balance*. Just as for hop-derived characters, so too must parameters such as sweetness-bitterness, volatiles, and so on be balanced.

The myriad of interactions that may take place in the human taste and olfactory system following the consumption of beer is enormously complicated. Certainly there is only limited knowledge of the physiological basis for them. Yet it is perfectly possible, indeed essential, for the Brewer to design products that delight the consumer because their flavor characteristics are so carefully balanced, with or without high overall flavor impact. These are the products that a consumer will find drinkable and will be tempted to order again. Overtly flavored products undoubtedly interest, and are enjoyed by, some consumers, but the biggest-selling brands worldwide tend not to have extremes of taste or aroma.

MOUTHFEEL. One of the least understood aspects of beer flavor is mouthfeel, which is sometimes referred to as body or texture. It seems unlikely that the perception of body relates to one or even a very few components of beer.

Recently a vocabulary to describe what expert tasters perceive as mouthfeel has been described. One of the terms is "tingle," which is quite clearly directly related to the carbon dioxide content of a beer. CO_2 reacts with pain receptors in the palate (leave your tongue in a highly carbonated beer for a few minutes and you'll see what I mean), and yet most people find this sensation to be pleasurable. For many beers, a relatively high concentration of CO_2 is essential to deliver this effect, which is due to an influence on the trigeminal nerve. In the United States, most beers are relatively highly carbonated (in excess of 2.6 volumes CO_2 per volume of beer). However, English ales traditionally have been of low carbon dioxide content, and a new genre of low-CO_2 keg ales has sprung up in recent years. These are the "nitrokegs," so named because the beer also contains nitrogen, both to support foam qualities and to impart the textural smoothness long known to be associated with use of this gas. The downside for some Brewers is that the use of nitrogen suppresses hoppy nose—one more example of how one aspect of beer quality influences another.

Nobody is certain of all the chemical species in beer that might influence texture. Some say the long and wobbly polysaccharides originating in the cell walls of barley have a role. Certainly they will increase viscosity, and some people suspect that increased viscosity is an important contributor to mouthfeel, as it will alter the flow characteristics of saliva in the mouth. Others have championed proteins, chloride, glycerol, organic acids (such as citric acid and acetic acid), and even ethanol itself as determinants of mouthfeel. Some believe that the polyphenols that we referred to earlier are important. They have long been known to cause astringency in ciders, and astringency is certainly one term in the mouthfeel vocabulary, but the levels of tannin found in beer are substantially lower than those in red wine or hard cider.

FLAVOR STABILITY. While flavor is conveniently described in terms of the individual components of a freshly packaged beer, the quality of beer most definitely changes over time. Such changes will be much more readily apparent in the more subtly flavored lager-style products. The nature of the changes differs between beers, but will generally include a decrease in bitterness and increase in sweetness, the development and subsequent decline (thank goodness!) of a ribes character (blackcurrant buds, tomcat pee), and the development of a cardboard or wet-paper note, followed by winey, woody, and sherry-like characteristics.

Despite years of extensive research, there is no consensus among brewing scientists regarding the origin of all the chemical species that cause this character. Some champion the bitter compounds, the iso-α-acids, as being a prime source. Others believe that certain alcohols that are produced by yeast during fermentation are important. The majority believe (rightly or wrongly) that the staling of beer, like that of other foodstuffs, can be traced to the oxidation of unsaturated fatty acids that originate in the malt.

What is for sure is that the degradations that lead ultimately to the development of staleness depend on the presence of oxygen. For this reason it is essential during packaging to minimize the ingress of oxygen into the can, bottle, or keg. Furthermore, oxygen uptake into the package in trade must be avoided. This is only really significant in bottled beers, where oxygen can sneak in through the seal between the crown cork and the neck of the bottle. Recently some Brewers have used corks that have an oxygen scavenger fused into them.

Despite the precautions taken to avoid oxygen access to beer in the package, all beers stale eventually. There is an increasing conviction that the tendency to form this cardboard character is built into the product during the production process. And so Brewers have started to consider eradicating oxygen uptake throughout the brewery; they have even started to suspect that the oxidation reaction has already started in the malting operation. (An unwritten rule for too many brewers is "pass the buck and blame the maltster"—and as often as not this is unjust.) The reason why these carbonyl compounds don't reveal themselves during the process may be that oxidation only goes as far as an intermediate that subsequently breaks down in the package over time. Alternatively, it may be that the staling compounds are produced early in the process and bind onto other compounds (principally sulfur dioxide, which is a natural product of fermentation, and also amino acids) and the complexes formed progressively degrade in the beer, exposing the carbonyl character.

Like other chemical processes, the staling reaction is retarded at reduced temperatures. A can of beer on an unrefrigerated shelf in Death Valley will stale in a couple of days, whereas the same beer in an icebox will still be fresh six months later. In some markets, notably the States, refrigerated distribution is widely employed.

DRINKABILITY

The drinkability sessions in Burton were indeed popular. Starting around 5 P.M., you'd start with half-pint glasses of three beers and the instruction to comment on them before choosing the one you would like to continue with. Having made your selection, you would be given a further half pint and off you went to do your thing—play cards, throw darts, or just talk. Whenever the need took you, off you headed for a refill of whichever of the three beers you wanted. Each time, your choice was noted down and a tally kept of the pints devoured from each keg or cask. At the end of the evening you were poured into a cab, clutching a questionnaire for your spouse or partner to fill in concerning the impact that your evening's drinking had had in various behavioral categories that we need not go into here.

We were looking for factor M, the moreishness or drinkability component of beers that we believed might fairly explain why it is that some beers are inherently more quaffable than others. Alas, we never found it.

Several of us were variously convinced that the factor was a nucleotide, or salt levels, or an amino acid such as glutamate (a thought triggered, no doubt, by the well-known role of MSG in making Chinese food impossible to stop nibbling at).

Nothing we tried seemed to pan out. Just about the only firm conclusion we ever drew was that excessively bitter and aromatic brews are less drinkable than more modestly hopped ones.

Drinkability means different things to different people. For me it is a case of "that was a good pint; you know, I could really drink another of those." For others it all has to do with satiety. "I really couldn't take in another of those," whether it is because they are filled (with fluid or carbon dioxide) or whether it's a case of "mmm, that is so fully flavored that my senses are saturated," a situation that surely pertains with some of the outrageously bitter and hoppy aromatic North American microbrews.

Mark Conner of the Centre for Decision Research at the University of Leeds describes four separate components of satiety. The first ("specific sensory satiety") is the sensory response per se: how the characteristics of a beer—its flavor, mouthfeel (in which carbonation and nitrogenation have a key role to play), and temperature—impinge upon one's feeling of fullness. Appearance may also have a psychological impact—how hard is it to savor a cloudy beer if you know that it is one that is supposed to be bright?

The visual appeal may literally get our digestive juices flowing and thirst response activated. Just look at the imagery of beer in advertisements. Depending on your conditioning, what does the appearance of a cold one do for you? A beer with foam is inherently more appealing than is one without. We must not discount the other components of the sensory experience, either: lighting, locale, and the other stuff that we are consuming, whether another foodstuff or cigarettes, for example.

The second component pertains to an individual's belief about a beer. What does the imagery do for them? What prior experiences have they had with this beer, both pleasurable and otherwise? What have they *heard* about a given brand? There are some beers with unfounded reputations for causing headaches and hangovers; such urban mythology marches against drinkability. In this category we might also include other psychological issues—such as messages about a beer's healthfulness, or the imagery associated with it. Is it good to be seen with this bottle of beer?

The third component of the satiety effect (with beers consumed in moderation) is bodily responses such as distension of the stomach wall and the rate at which the stomach empties.

The fourth element is the direct impact of the components of the beer and its digestion products as they course through the blood and hit the brain.

Another way to view drinkability of a beer is in the context of the extent to which it is thirst-quenching. My colleague Jean-Xavier Guinard found a positive correlation of drinkability with carbonation and bubble density. Negative determinants included overall aroma and flavor, color, viscosity, malty, hoppy, burnt, bitterness, acidic, metallic, astringency, and aftertaste. In other words, and entirely as one might predict, the beers predicted to be the easiest to drink were those of "gentlest" flavor. If surprised about this, consider the beers worldwide that sell the largest volumes.

Surely it is all about balance? A relatively bland beer can be just as drinkable as a deeply flavored one, as long as they are both in balance. No overt hoppiness, unless tempered with malt and alcohol.

ASSESSING DRINKABILITY. Testing for drinkability is a real challenge, on the grounds of relevance but also ethics. One approach is to ask people to rank beers at different stages of drinking: after the first sip and then after finishing half, one, and one-and-a–half liters. In this way it is possible to see if opinions of a product's acceptability changed as more was consumed. Another way is to measure people's thirst after consumption of beer by totting up how much water is sipped between taking a drink of beer every 15 minutes.

Fushiki and colleagues from Kyoto University measured stomach volume and correlated it with sensory attributes of beer. After a controlled rate of beer intake, the subjects were told to urinate every half hour over a period of 3 hours and some unfortunate lab technician was charged with the task of measuring the volume collected. The subjects were also asked to give an opinion on how tasty they found the beers, whether they were keen on drinking them, how full they felt their stomachs were, and so on. The conclusion was that the least drinkable beer was the one that resided in the stomach the longest. They later showed that the less fresh the beer, the less drinkable it is and the less is its tendency to make you want to urinate.

In summary, I would suggest that it really is excess of any component, including carbonation, that induces satiety. However, a high carbonation, with its attack on the trigeminal pain response, is an exquisite sensation. Try it: next time you are thirsty have some warm tap water or some cold carbonated spring water and see which one scrapes the back of your throat and livens up your taste buds more effectively.

So it is for beer, too: a cold, highly carbonated canned lager works better to refresh your palate than does a pint of traditional ale drawn straight from the barrel. But which one is the more drinkable, in terms of moreishness? Provided it has not turned to vinegar, is not over-hopped, and does not have unappealing lumps that have evaded the fining process, then for me it is the ale.

BEER AS A FOODSTUFF

IS BEER GOOD FOR YOU?

For many years Guinness advertised their beer with the slogan *Guinness is good for you,* before changes in law decreed that this type of claim could no longer be made. Later, also in the British Isles, came claims (less overt ones perhaps) for another beer: *A Double Diamond works wonders.* A Shakespearean actor, Sir Bernard Miles, was featured in a television campaign extolling the virtues of Mackeson Stout, using the immortal lines: *It looks good, it tastes good, and by golly it does you good!* That beer was one of the stouts served in the maternity wards of British hospitals. Now, in the early twenty-first century, no longer does it seem to be politically correct to make claims that beer drinking is good for you, despite the growing scientific evidence in support of such statements. Beer truly is "liquid bread"—and rather more besides.

Of course, a broad spectrum of opinion exists concerning the desirability or otherwise of consuming alcoholic beverages. For millions worldwide, such drinks are forbidden on religious grounds. Among cultures where alcohol is tolerated, right-minded individuals recognize the social unacceptability of consuming alcohol to excess, with the terrible price it can have for some through road traffic accidents and family distress through alcoholism and for others through the development of conditions such as cirrhosis of the liver and certain types of cancer. This is quite apart from the impairment of performance that drinking at inappropriate times can cause. Increasingly, however, it is becoming recognized that there may be some health benefits associated with the consumption of alcoholic beverages in moderation—and not only by helping to reduce stress and stress-related problems such as increased excitability and heart rate.

The health-related recognition of beer stems back to Ancient Egypt. Beer was used as a mouthwash, enema, vaginal douche, and applicant to wounds, quite apart from its importance as a key component of the diet. It seems that on Captain Cook's ships beer contributed as many calories to the sailors' diets as biscuits and meat combined. Perhaps John Taylor, who kept an alehouse in London, England, penned the most glowing reference to the benefits of drinking ale. In 1651 he suggested that ale

> is a singular remedy against all melancholic diseases, *Tremor cordis*, and maladies of the spleen; it is purgative and of great operation against *Iliaca passio*, and all gripings of the small guts; it cures the stone in the bladder, reines or kidneys and provokes urine wonderfully, it mollifies tumors and swellings on the body, and is very predominant in opening the obstructions of the liver. It is

most effectual for clearing of the sight, being applied outwardly it assuageth the unsufferable pain of the Gout called *Artichicha Podagra* or *Gonogra*, the yeast or barm being laid hot to the part pained, in which way it is easeful to all impostumes, or the pain in the hip called *Sciatica passio*...and being buttered (as our Gallenists well observe) it is good against all contagious diseases, fevers, agues, rheums, coughs and catarrhs.

That's quite a testimony. We can say, though, with rather more careful consideration and supportive evidence, that there are indeed potentially positive aspects to the drinking of beer.

In comparison with other alcoholic beverages, the content of alcohol is relatively low in the majority of beers. The alcohol strength of beers, which for the most part tends to be in the range 3%–6% by volume, is much lower than that of most other alcoholic drinks. Beer, then, is more suited to the quenching of thirst and counteraction of dehydration than is wine, for instance. In some countries (such as Germany) beers at the lower alcohol end of the spectrum are favored as sports drinks, because their osmotic pressure is similar to ("isotonic with") that of body fluids. Such beers do possess some calorific value as an energy source, because they do contain some carbohydrate as well as ethanol. Incidentally, all beers are fat-free.

It has been claimed that beer (though not the alcohol within it) stimulates milk production in nursing mothers and may reduce the risk of gallstones and promote bowel function; it has even been claimed in Japan that materials produced during the kilning of malt and that enter into darker beers suppress the onset of dental caries. There is good evidence that beer in moderation cuts down the risk of diabetes mellitus.

Beer contains some vitamins, notably some of those in the B group (pyridoxine, niacin, riboflavin, and folic acid) and minerals, especially magnesium, potassium, and selenium. Beers generally have a low ratio of sodium to potassium, which is beneficial for blood pressure. There are usually quite high concentrations of calcium and phosphate in beer, and also of silica, which supposedly promotes the excretion of potentially harmful aluminum from the body (aluminum being one of the purported causative agents in Alzheimer's disease), as well as contributing favorably to bone structure and the countering of osteoporosis. Indeed beer, alongside bananas, is the richest source of silica in the diet.

Recently a number of publications have drawn attention to the importance of antioxidants in foodstuffs and the possible contribution that these could make to the diet in terms of protecting against oxygen radicals, which are understood to have undesirable influences on the body. Most significant of these antioxidants are the polyphenols, which are present, inter alia, in beer. They may also include ferulic acid. Incidentally, in the case of ferulic acid (and also silicon), it has been demonstrated that it is "bioavailable"—that is, it actually does get into the body and, presumably, gets to do the good that is claimed for it. All too often it is not clear that the supposedly beneficial constituents of foods even get assimilated into the body. The plethora of work on red wine highlighting its very high content of polyphenols has seldom extended to confirming a simple correlation between concentration of these molecules and the benefit to the body, that is, the higher the level in the drink, the better it is. Might it be that all that is needed is a certain quantity, and, beyond a certain amount, excess is irrelevant (or even detrimental?) The analogy would be filling a car with gasoline. If you only need to travel ten miles, then a tank full of gas is pointless.

Consideration of all these benefits from drinking beer makes it small wonder that, for generations, stouts were a recommended aspect of the diet of nursing mothers. These days, of course, one is much more likely to find pregnant women dissuaded from consuming alcohol in all forms, for risk of harm to the unborn infant.

Most debate in recent years has focused on the relative merits and demerits of consuming ethanol itself for the broader adult populace. Evidence emerging from the medical community that moderate drinking correlates with lower death rates due to various causes led in 1987 to the U.K. government raising its recommended maximum for drinking by adults: men are advised to drink no more than 21 units a week (a unit in the United Kingdom is 8 g of alcohol, which is roughly equal to half a pint of medium-strength beer) and women 14 units. Additionally, the advice is to consume no more than 4 units per day. Compare this with the recommendations of the French, whose more liberal attitude to alcohol and predilection for wine prompts them to advise men to drink no more than one bottle of wine per day, and women half as much.

In particular, there seems to be evidence for alcohol protecting against cardiovascular disease. These effects may be linked in part to a component of beer other than alcohol itself, but ethanol appears to be key in altering the balance of the high- and low-density lipoproteins in the blood so that the deposition of fats on artery walls is reduced. Alcohol also appears to reduce the "stickiness" of blood platelets, making them less likely to aggregate together as blood clots.

Another component of beer that may have a hypocholesterolemic influence is the β-glucan. This is the principal component of the cell walls of barley, and it can cause all sorts of problems for the Brewer (see chapter 8). However, if this polysaccharide survives into beer it represents soluble fiber, which has been claimed to have a cholesterol-lowering effect.

Wide ranges of clinical studies have reached the conclusion that there is a "U-shaped" or "J-shaped" relationship between deaths and alcohol consumption (figure 4.2). Modest consumption of alcohol lowers the relative risk of death, particularly through a lesser incidence of coronary heart disease. This

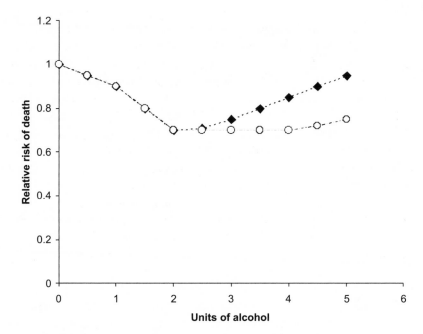

FIGURE 4.2 The impact of moderate alcohol consumption on mortality risk: circles depict risk of atherosclerosis; diamonds depict risk from all causes.

relationship appears to hold across national boundaries and cultures, and was most famously publicized as the so-called French paradox: a people famed for their enjoyment of fine food high in saturated fats leading to high levels of cholesterol in blood serum nonetheless reported some of the lowest frequencies of deaths from coronary heart disease. Hence we see a justification for their relatively high recommended alcohol intake, although it has also been suggested that it is not only the alcohol in drinks that has a beneficial effect but also other components of the so-called Mediterranean diet, such as garlic.

Some authors argue for a superior beneficial effect of alcohol when taken in the form of red wine rather than beer. It seems, however, that most if not all of these studies have failed to take into consideration confounding factors in the diet. Merely to compare wine drinkers with beer drinkers and to ignore the other elements of their respective lifestyles is misleading. Exaggerating to make the point, we might say that wine drinkers eat lettuce leaves, work out at the gym and occupy higher socioeconomic strata than do beer drinkers, who stuff burgers and watch ball games as couch potatoes. Incidentally, in 19th-century England, teetotalers were regarded as cranks who were jeopardizing their health by ignoring beer, which was considered to be a key feature of the diet.

In summary, it does very much seem that there may be benefits associated with the moderate consumption of alcoholic beverages, including beer, by mature and well-adjusted adults, at the appropriate time and in the appropriate place. Drinking to excess always has been and always will be antisocial, dangerous, and unacceptable. Back in 1789, James Madison expressed the wish that "the brewing industry would strike deep root in every state in the Union." The concern was with the growth in consumption of hard liquors—to the extent that in Massachusetts a law was introduced exempting breweries from excise taxes for five years. The legislature said that "The wholesome qualities of malt liquors greatly recommend them to general use, as an important means of preserving the health of the citizens of this Commonwealth."

How strange it is, then, that there still appear to be people seeking to push beer down in the eyes of the populace in comparison to other alcoholic drinks and, indeed, other components of the dinner table. For example, recent years saw the emergence of the South Beach Diet, which divided carbohydrates into "good carbs," which are slow to be digested, and "bad carbs," which head straight into the blood supply and are used rapidly to stoke up the body's larder and, well, make you fat. The author of the diet said that the worst food of all is beer, as it was loaded with maltose, the worst sugar of all. I wrote to that author, pointing out that for most beers the maltose is entirely removed through the action of fermentation. No reply—although later versions of the diet have rescinded the "ban" on beer due to "recent research."

The reality is that the majority of calories in a beer are contributed by alcohol itself. A 12-ounce serving of a 5% ABV North American lager amounts to some 150 calories, whereas a light beer is around 100 calories—much the same as a glass of wine. Beer and wine are very much of the same order of magnitude when it comes to calories on a "by serving" basis. There is absolutely no justification for the concept of a "beer belly." Sure, some folks who drink beer are overweight and many wine drinkers are skinny. It's lifestyle again: many beer drinkers enjoy a sedentary existence, whereas, if I look around in California, I see many trendy folks jogging as they push their prams, lead their dogs, and head home for a glass of pinot noir.

The carbohydrates that are present in beer tend to be "good carbs" in that they are mostly polymerized (shortish chains of linked glucoses), and it is these "slow release" (they are metabolized in a slower, less impactful way) and in some instances "no release" molecules that are claimed to be best for the body. Some of them can be considered as soluble fiber (they are similar to the stuff touted in oat-based breakfast cereals) and some of them are probably prebiotics, which are the substances recommended in food that head to your colon to feed beneficial microorganisms.

BOX 4.2 ALL ABOUT HEALTH

A beer label in the United States must list the brand, the class or type of product, the name and address of the production plant, the net contents, and, depending on the state, a declaration of alcohol content. It is also mandatory to post the following health warning:

1 According to the Surgeon General, women should not drink alcoholic beverages during pregnancy because of the risk of birth defects

2 Consumption of alcoholic beverages impairs your ability to drive a car or operate machinery, and may cause health problems

And yet the real merits of drinking beer in moderation are as recognized now as they were in the very first days of the United States. The early settlers were obliged to drink infected water, whereas if they could get hold of imported ale (or produce their own), their thirst was quenched with a much safer product. There is a boiling (i.e., sterilization) stage during brewing, added to the fact that the low pH and presence of the antiseptic bitter components from hops contribute to inherent microbiological robustness. No pathogens will grow in beer. And beer has always been the drink of moderation. President Jefferson said "I wish to see this beverage become common" because much stronger distilled products of dubious provenance were gaining sway.

The notion has arisen among too many that beer is simply "empty calories." Yet a study commissioned by Britain's Royal Society in 1939 said that a barrel of beer is the equivalent in cumulative nutritive value of 10 pounds of beef ribs, 8 pounds of shoulder mutton, 4 pounds of cheese, 20 pounds of potatoes, 1 pound of rump steak, 3 pounds of rabbit, 3 pounds of plaice, 8 pounds of bread, 3 pounds of butter, 6 pounds of chicken, and 19 eggs.

More insidious than the empty calorie lobby are the blinkered antialcohol campaigners who cannot countenance that there is any value whatsoever to be had from drinking alcoholic beverages, including beer. Their dogma and mantra is that to take one drink is to step onto the slippery slope that will lead to inevitable and unavoidable alcoholism. The neo-prohibitionists close their eyes to the unarguable fact that by far the majority of people who enjoy alcohol don't feel a compulsion to drink and don't suffer from withdrawal, which are the markers of an addictive drug. It is the likes of Stanton Peele (*The Meaning of Addiction,* Lexington, Mass.: Lexington Books, 1985) and Herbert Fingarette (*Heavy Drinking: The Myth of Alcoholism as a Disease,* Berkeley: University of California Press, 1988) who have driven a wedge through the myth of alcoholism as a disease. Alcoholism is a form of addiction, and those who abuse alcohol may also display other forms of compulsive behavior, such as addictions to drugs, cigarettes, gambling, shopping, sex, and caffeine.

There remains a forum of scientists and medics who are convinced that alcoholism is at least in part inherited through the genome. Studies with adoptive twins claim to show that children born to an alcoholic parent and put up for adoption soon after birth display an increased tendency toward alcoholism, as compared to those born to nonalcoholic parents. Careful analysis of such studies reveals that four-fifths of the adoptees that came from an alcoholic biological father did not become alcoholic, which either means they did not inherit (or express) the relevant gene(s) or that there are other impacting factors, notably environmental ones. As yet, no consistent "trait marker" for alcoholism has been identified.

Many are convinced that the drinking of alcohol, probably beer more than beverages such as wine, cannot be divorced from irresponsible behavior. In my native Britain, websites at the highest level are highlighting concerns about binge drinking, especially among the young (www.pm.gov.uk/output/page4501.asp.) Is irresponsible advertising to blame for the perceived epidemic of irresponsible drinking? The prime minister's site stresses the cost to the health service accruing from excessive alcohol consumption, but there is no attempt to quantify the cost *benefit* in terms of reduced costs achieved by keeping people out of hospitals and out of the graveyard by their taking alcohol in moderation, with the attendant countering of atherosclerosis. A. Britton et al., "A Comparison of the Alcohol-Attributable Mortality in Four European Countries" (*Eur J Epidemiol,* 18 [2003]: 643), conclude that there are 2% fewer deaths annually in England and Wales than would be expected in a nondrinking population.

Nobody should ever downplay the risks and concerns of excessive drinking, especially in the context of a decreased ability to operate a machine such as an automobile. Yet there are complex social and attitudinal factors at play here, too. Alcohol seems to feature in 15% of fatal crashes in the United Kingdom, where the legal drinking age is 18, but more than 30% in the United States, where the legal drinking age is 21.

I am an Englishman living in California and I have to say that my country of residence has a far more hypocritical, naive, and illogical stance when it comes to alcohol than has the land of my birth. The younger element in the United States, fed scare stories and one-sided arguments about alcohol, too often perceives drinking to be some mysterious and hidden pleasure, a "forbidden fruit." Perhaps it is not to be wondered at that when students reach the legal drinking age of 21 they often succumb to the temptations of the drinking ritual (usually involving the taking of 21 shots of spirit), too often with devastating consequences. The key word is education—the teaching of sense and moderation in all matters, not least alcohol. As Dwight B. Heath said (in *Controversies in the Addictions Field,* ed. Ruth C. Engs, 1990, American Council on Alcoholism, Dubuque), the reinforcement of standards by family and friends will be more effective than legal and regulatory controls. Alcohol is not something to be swept under the carpet, but something positioned as a wholesome and worthy component of a healthy, fulfilling, and responsible lifestyle when consumed in moderation.

And what is moderation? This is tremendously difficult to quantify and has deep cultural grounding. A family in France or Italy, for example, would never consider the regular

(continued)

BOX 4.2 CONTINUED

consumption of bottles of wine with dinner as being immoderate, any more than would a party of Prague postmen enjoying liters of lager. The World Health Organization suggests that 60 grams of alcohol per day should be a maximum. For a beer of 5% alcohol by volume, which equates to approximately 4% alcohol by weight, this means 1.5 liters.

WHY BEER CAN BE SAFER TO DRINK THAN WATER

Beer is most inhospitable to the growth of microorganisms. The boiling stage in beer production kills the vast majority of organisms that might have entered into the process. During fermentation the pH falls to about 4.0, which is too low for most organisms to thrive, and of course most of the nutrients that a contaminating microbe would need are efficiently consumed by yeast. At the same time, ethanol is produced, which itself protects against microbial growth. Most beer is packaged under relatively anaerobic conditions, preventing the growth of any microbe that requires oxygen. And it has been proven that those pathogenic bacteria that don't require oxygen are unable to populate beer. Above all, beer contains various substances that suppress bacterial growth. These include some of the tannins, but in particular it is the bitter compounds, the iso-α-acids, which have a profound antimicrobial influence. (What wonderful things these substances are: they make beer nicely bitter, they help provide the foam, and they prevent bug infections—what a pity they are the cause of skunky flavor!)

BOX 4.3 BEER AND SEX

I was once sitting with a man (who shall remain nameless) beneath a glorious sun in a brewery garden in South Africa. We were at a convention, and the hosts had recruited the most beautiful girls to serve beer for the occasion. Now my friend could hardly be described as an oil painting. When good looks were handed out, he must have dozed off. Well, actually, he must have been in the deepest possible slumber. He didn't smile very often, either, and wasn't given to oratory. But as we sat enjoying our brews and admiring the scenery, he finally turned to me and said, "Have you noticed, Charlie, that the more beer you drink the more attractive you become to women?"

How very prescient. And thoughts of that balmy day more than a decade ago set me to wonder about the impact that our favorite drink has had in the success or otherwise of man and woman's sexuality. It's an old joke that beer has been helping ugly people have sex for generations—but is it actually a help or a hindrance?

As Shakespeare wrote in *Macbeth*, "it provokes the desire, but it takes away the performance."

This draws attention to two disparate impacts of alcohol on sexual performance (and we must perforce generalize—for there is nothing unique about beer in this respect; the same goes for wine and spirits): on the one hand, its effects on the mind, and on the other, those on other bits of the body.

Alcohol seems to lower the production of certain hormones, but only when consumed in substantial quantities. Men drinking less than 3.5 standard drinks per day had perfectly normal sperm. Meanwhile, estrogen levels are increased in pre-menopausal women taking up to 2 drinks per day and in post-menopausal women partaking of no more than one drink per day. This may be one of the reasons why alcohol consumption helps counter osteoporosis.

There is evidence that alcohol consumption in a dating situation triggers certain beliefs, even in those people who are drinking with genuine moderation. One study showed that women drinking an alcoholic beverage were perceived by men as being less attractive and socially skilled than those partaking of soft drinks. Meanwhile, socially anxious men displayed impairments in interpersonal skills in mixed sex situations after drinking alcohol.

Alcohol also impaired the perception by college men of negative feedback from females—shades of my friend in the South African garden. In yet another study, it was found that women consuming alcohol found their male partners more likeable—and the men in turn felt that their drinking female friends found them more likeable. When both were drinking, they felt their partner to be more extroverted.

All of this suggests to me that the concept of a beer overtly marketed as a sexual promoter is, well, rather silly. Of course, it has been done. Back in 2000, a beer called Rethink came out of British Columbia, one that contained herbs claimed to increase sexual performance: ginseng, gingko, and tribulus. They put a number on each beer cap corresponding to a sexual position in the Kama Sutra, and all you had to do was "pick your poison," check out the number, and head to the website to find out what gymnastics were recommended for you and your friend. Sounds to me less like Rethink than Think Again. In fact, I just went to their website and can find no mention of sex.

A year or two ago, some good friends in the industry had me try their new beer, which is not overtly marketed on the basis of any beneficial impact on "matrimonial Olympics" but does contain ginseng, guarana, and caffeine. I swear that after one sip I felt as if I would stay awake for a week. They looked at me and said "It's not for you, Charlie."

BEER STRENGTH: ITS RELEVANCE

The strength of a beer is defined by its alcohol content. At one extreme can be found beers containing more than 13% alcohol by volume (ABV). At the other pole are the alcohol-free beers. Quite apart from the obvious variation in physiological impact that beers of different alcoholic strength will have, the alcohol

TABLE 4.1. TYPICAL ALCOHOLIC STRENGTHS OF VARIOUS BEVERAGES

Drink	Typical alcohol content (% ABV)	Volume of drink constituting a "unit"
Premium beer	4.5	approx. half a pint
High-strength beer	9.0	approx. quarter pint
Wine	12.0	approx. tenth of a 75-cl bottle
Whiskey	40.0	20 ml
Gin	40.0	20 ml
Vodka	45.0	15–20 ml
Vermouth	15.0	approx. 1/15 of a bottle

influences flavor directly and indirectly as well as the foaming properties of beer, as we have seen.

Perhaps of most importance, though, is the fact that in many countries, the tax payable is levied on the basis of alcoholic strength. This is not the case in the United States, where fixed-rate levies are made at federal and state levels on a per barrel basis, irrespective of the strength of the beer in the container (see chapter 1).

In the United Kingdom, however, the amount of duty levied is in direct proportion to the alcohol content of the beer (see chapter 11). Small wonder that the precision with which alcohol content of beer can be measured is most important, both to the Brewer and to HM Revenue & Customs. Indeed, in all countries, it is most important that careful checks and records are made of volumes of beer, because that always has a direct impact on the size of the check that the Brewer will be writing to the taxman.

Typical alcohol levels for a range of alcoholic beverages are shown in table 4.1, together with the volume of that drink that constitutes a "unit."

We have discovered in this chapter that a vast myriad of compounds and physical interactions influence the quality of beer. Let just one of them be out of balance, and the whole product will be ruined. Time now, then, to walk steadily through the malting and brewing processes to see how it is that the devoted Maltster and Brewer strive to ensure that the balance in your beer is indeed right, time after time after time.

THE HEART AND SOUL OF BEER

MALT

More than 90% of the beer brewed worldwide has barley malt as the key grist component. True, some beers, such as the weissbiers in Germany, are produced from malted wheat and some long-standing beers in South Africa are based on sorghum. Interest in sorghum is increasingly global, for it forms the base of several beers designed for consumption by celiac sufferers who are sensitive to wheat and possibly barley proteins.

Many malt-based beers contain other grist materials, occasionally for reasons of cost but usually because they may introduce distinctive characters. Thus a major international brand features rice in its recipe, and some Brewers will use wheat-based adjuncts because they feel they enhance foam quality. These so-called adjuncts will be considered in chapter 8.

It is malted barley, however, that remains the foundation of most beers, and it seldom accounts for less than 50% of the grist. Frequently it will comprise the sole source of fermentable carbohydrate. Efficient brewing and top-quality beer are inextricably linked to the quality of the malt. This focuses attention on the quality of the barley and on the malting operation, both of which must be right.

BARLEY

Cultivated barley (*Hordeum vulgare*) belongs to the grass family and is grown in more extremes of climate than any other cereal (table 5.1). It has been estimated that barley emerged from its ancestor in Egypt some 20,000 years ago. Worldwide production of barley is now in excess of 138 million tons, but less than 25% of that is malted for the brewing of beer.

TABLE 5.1. PRODUCTION OF BARLEY

Country	Barley grown (million tons, 2005)
Russia and the countries of the former USSR	16.7
Canada	12.1
Germany	11.7
France	10.4
Ukraine	9.3
Turkey	9.0
Australia	6.6
U.K.	5.5
U.S.	4.6
Spain	4.4

Two types of barley are used for malting and brewing (figure 5.1). In two-rowed barley, two rows of kernels develop, one on either side of the ear. Six-rowed barley has three corns on either side of the ear. Space is restricted on the ear, meaning that some of the corns in the latter type must be twisted in order to fit. Six-row barleys tend to have a higher ratio of husk to starch, for the simple reason that the relative crowding on the ear renders less space available in their endosperms for starch to accumulate. On this basis, there is less potential fermentable material from six-row barleys, and therefore, all things being equal, a brewer (and therefore maltster) prefers two-row barley. However, six-row barley accumulates proportionately more protein and, as enzymes are proteins, these barleys afford higher levels of enzymes, making them better suited to use as part of grists that contain large amounts of adjuncts that of themselves do not possess enzymes.

The barleycorn consists of a baby plant (embryo) together with an associated food reserve (starchy endosperm), packed within protective layers (see figure 5.2). It is the food reserve that is of primary interest to the Brewer (and therefore the Maltster), as this is the origin of the fermentable material that will subsequently be converted into beer. The reserve comprises starch in the form of large and small granules packed within a matrix of protein, and the whole is wrapped up in a relatively thin cell wall (figure 5.2). The cells of the starchy endosperm (barleycorns contain approximately a quarter of a million of them) are dead, and, although they contain a few enzymes, most of the significant enzymes that are necessary to digest the food reserves can't be made by the starchy endosperm.

It is the embryo that has ultimate control on the breakdown of the endosperm. After all, this is *its* food reserve, the true function of which is as a

FIGURE 5.1 Two-row and six-row barley (courtesy of National Institute of Agricultural Botany)

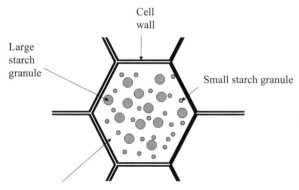

FIGURE 5.2 Diagrammatical representation of one cell of the starchy endosperm of barley. Only a handful of the numerous large and small starch granules are depicted.

source of nourishment to support its growth. The endosperm wasn't designed to oblige the Maltster or Brewer! The skilled maltster takes advantage of the embryo's ability to mobilize its food store to enable her to furnish the brewer with good-quality malt.

Enzyme production is the preserve of the aleurone tissue, which is four or fewer cells deep and surrounds the starchy endosperm. The embryo produces a series of hormones, which migrate to the aleurone, there to control the switching on or off of enzyme synthesis. Before an embryo can get going to produce these hormones and before any enzyme can act to hydrolyze the starchy endosperm, the barley must have its moisture content increased. Whole barley from a malting store will typically contain some 10%–13% water, with the embryo holding 18%–20% moisture. To commence metabolism of the embryo, this moisture content must be increased. The starchy endosperm, too, must be hydrated, for enzymes act more rapidly if their substrates are solvated.

In the malting process, then, barley is first steeped in water, to bring it up to a moisture content in the region 42%–46%. This triggers the synthesis and migration of enzymes into the starchy endosperm. The first enzymes produced are those that open up the cell walls and hydrolyze their constituents. Following on from these are the proteolytic enzymes, and the last enzymes to be made are the amylases, which are responsible for degrading the starch.

A major requirement in the production of malt for brewing is comprehensive hydrolysis of the cell walls, which leads to a softening of the grain and its easier milling and extraction. Secondly, there needs to be a substantial breakdown of protein, to free the starch from enclosure and to eliminate potential haze-forming material, but primarily to produce amino acids that the yeast will require as building blocks to make its own proteins and therefore grow. What the brewer does not want is significant degradation of the starch, for it is this that he wants to break down in the brewery to yield fermentable sugars.

The germination of barley, therefore, is carried out for a time long enough for cell walls and protein to be degraded and for starch-degrading enzymes to be synthesized, but not so long as to lead to excessive growth of the embryo. The process is generally referred to as "modification." Once modification is completed to the required extent, it is halted by kilning.

WHICH BARLEY?

Any barley, providing it is living (viable), can be malted, but the quality of the malt and the efficiency of the malting operation depend greatly on the nature of the barley. In turn, this makes certain demands of the farmer.

Barleys can be divided into so-called malting and nonmalting (or feed) grades. The division is based on the amount of extractable material that can be obtained from their malts in a brewing operation. Malting varieties give high levels of extractable material, whereas nonmalting grades don't. Or, more accurately, they don't when malted for conventional periods of time and brewed normally.

The difference between barleys lies in the ease with which their endosperms can be modified during germination. In nonmalting grades substantial areas of the endosperm will remain intact after conventional periods of germination (four to six days). This may be because water doesn't get distributed evenly throughout the endosperm, which in turn must have something to do with their structure. Alternatively, their cell walls may be less easily degraded than those of better grades, or they may be less capable of synthesizing enzymes.

Evidence suggests that all of these factors may be important. Feed grade barleys have a relatively high proportion of corns that have "steely" endosperms, in which the components are very tightly packed, meaning that neither water nor enzymes can easily gain access. Malting barleys, however, have more "mealy" endosperms, which distribute water easily and are readily accessed by enzymes.

Steely endosperms *can* be hydrolyzed—it merely takes longer than for mealy grain. Like the Brewer, the Maltster works to tight time frames and will ordinarily select barley varieties of the higher malting grades. The Brewer in turn is likely to insist on certain varieties. Given the choice of two malts that possess identical analyses, many brewers will opt for a variety they know, because unexpected problems might occur with unproven varieties.

New varieties are continually coming into the marketplace as the end product of plant breeding and, with similar rapidity, older varieties disappear. Each variety has its own name, and some rather colorful ones at that. For many years varieties such as Plumage Archer, Maris Otter, and Proctor led the way in the United Kingdom; they were relatively long-lived in that they were used year after year by Brewers who were convinced of their importance in the brewing of top-quality ales. Indeed, there are still a few Brewers who swear by Maris Otter today—and malts made from this variety are very popular with the microbrewers in the United States. These days you'll find many more varieties across the world.

For it to be accepted, a newly bred variety must be demonstrably different from an existing cultivar and possess some advantage, for instance higher extractability. Not only must a new malting variety be capable of performing well in the malting and brewery, it must also possess the necessary properties when growing in the field, such as high yield, disease resistance, relatively short stiff straw of uniform length, and early ripening.

To be accepted for malting, any barley will also need to satisfy other criteria. First of all, it should have a relatively low content of protein. For a given size of grain it is self-evident that the more protein packed within it, the less room there is for other components. In other words, the more protein, the less starch, and it is, of course, the starch in which the Brewer is primarily interested, rather than a high protein content, which is needed in feed barley. Generally speaking, protein levels will be lower in grain from barley grown on lighter soils. Maltsters place a specification on barley for its protein content, and this in turn obliges the farmer to restrict the use of nitrogenous fertilizer. Accordingly, yields of malting barley tend to be lower than those of crops grown for feed purposes. To compensate for this, a malting premium is paid for barley that meets the necessary criteria of uniformity in variety and low protein.

Other specifications must be met, too. First and foremost, the barley must be living. If the embryo has been killed—which can occur all too easily, for example, if the barley has been badly dried—then it is incapable of producing the hormones that promote germination. Viability can be quickly checked by a staining test: living embryos cause a colorless dye to turn red.

Even if an embryo is alive, it may still not be capable of immediate germination. This is the so-called dormant state, and it is normal, albeit an irritation to the maltster. Quite what controls dormancy in plant seeds is not absolutely known. It tends to vary from variety to variety, and it also depends on environmental factors. The further north barley is grown, the more it tends to display dormancy. Cool and wet conditions in the growing season promote dormancy. The phenomenon might be an irritation to the Maltster, but it's important to the barley! If dormancy didn't exist, then the grain would germinate prematurely on the ear and not at the appropriate stage when it had left the parent plant and found its own bit of Mother Earth in which to sprout. It is for this same reason that the phenomenon is actually important to the Maltster, too. In certain climatic conditions, such as high rainfall at certain stages in the growing season, grain can start to chit (sprout) on the ear. When the barley is harvested and dried, the heat kills the growing embryo and the malting process is jeopardized.

Dormant barley must be stored to allow it to recover from this condition. Various treatments have been recommended for the release of barley from dormancy, including warm storage (e.g., 30°C/86°F) or, ironically, cold storage! It is certainly the case that a Maltster might almost welcome dormancy if it was a condition that he had total control over and could switch on or off at will in order to optimize barley purchasing and turnover.

A phenomenon related to dormancy is water-sensitivity. All barleys to a greater or lesser extent display this trait, in which germination is inhibited by the presence of too much water. There are various plausible explanations for the effect, the most likely being the role of water in inhibiting the access of oxygen, which the embryo requires to support respiration. Recognition of the phenomenon nearly sixty years ago led directly to the introduction of interrupted steeping regimes in the malthouse. Previously barley had been steeped continuously in water. Once it was realized that this would suppress embryo activity by "swamping" it, procedures were introduced whereby barley is steeped for a shorter period of time, followed by a draining stage and ensuing "air rest." Then more steep water is applied, followed by another air rest, and so on. The precise regime is optimized for each variety, but the process seldom takes longer than 48 hours, whereas before the days of interrupted steeping it took at least twice as long.

Brewers prefer barleys with larger corns, as they possess a larger ratio of starch to peripheral tissue (e.g., husk, which can amount to as much as 10% of the weight of the grain).

The reader might suspect that one barley looks very much like another and it would be difficult to tell them apart. True, it is difficult, but an expert will be able to inspect a handful of grain and pretty much be able to identify the variety by studying things like the color of the aleurone (some are white, others are blue), the size of the corns, and how wrinkly the hull is. Provided the sample is representative of the entire shipment, she will be able to tell whether she is looking at just one variety or a mixture. As barley varieties differ substantially in their performance, it is vital that they be malted separately, and the buyer would be entirely justified in rejecting a batch of barley on the basis of visual assessment alone. If further evidence is warranted, this can be obtained by a protein fingerprint: the proteins of the grain are extracted and separated on gels across which an electric current is passed. The proteins migrate to different extents on the gels before they are detected by staining. The patterns obtained are a characteristic of the variety. More recently, just as in forensic science, the analogous DNA fingerprinting has been suggested for barley; one might say it's to detect the crime of fraud in barley sales. A visual inspection, though, is generally sufficient—and it delivers, too, a verdict on whether the barley is free from infection and physical damage. It is rapid and can be performed when the barley is taken into the malthouse.

Finally, different barleys possess different inherent flavors. Increasingly, barleys are selected on the basis of an absence of undesirable flavor notes.

COMMERCIAL MALTING OPERATIONS

There are four basic process stages in a modern malting operation: intake, drying, and storage of barley; steeping; germination; and kilning.

INTAKE, DRYING, AND STORAGE OF BARLEY

Barley can be classified into two categories depending on when it is sown and harvested: winter varieties are sown in the fall, whereas spring varieties are sown in spring and harvested a little later than the winter varieties. Generally speaking, the earlier in the year the seed is sown, the lower will be the protein content in harvested grain (and the higher its yield), because starch accumulates right through the growing season.

Purchase of grain by the Maltster will be according to an agreed specification, which will include freedom from infection and infestation, protein content, grain size, viability, and moisture content. The farmer will be paid proportionately less for batches of higher water content because the Maltster will be obliged to dry them to an increased extent to prevent spoilage by insects and microorganisms. Most Maltsters don't favor the farmer taking responsibility for drying, for fear of destroying the embryo. In many parts of the world, including North America, drying is unnecessary because the barley is harvested with a sufficiently low moisture content (12% or lower).

The grain will arrive at the malthouse by road or rail, and as the transport waits it will be weighed and a sample will be tested for the key parameters of viability, protein content, and moisture. Expert evaluation will also provide a view on how clean the sample is in terms of weed content and whether the grain "smells sweet." A few grains may be sliced in half lengthwise and their endosperms assessed as to whether they are mealy or steely. Remember, the Maltster prefers mealy endosperms. Once accepted, the barley will normally be cleaned, to remove everything from dust and weeds to dead rodents, and screened, to remove small grain and dust, before passing into a silo, perhaps via a drying operation.

It is essential that in storage the grain be protected from the elements, yet it must also be ventilated, because barley, like other cereals, is somewhat attractive to microorganisms, insects, and animals. The risk from different pests and diseases differs tremendously between sites and environments. Frequently, no protective agents need to be employed, and they won't be unless it is absolutely necessary, but it is essential that pockets of infection by molds and fungi not occur and that the site be free from infestation by insects and vermin.

Barley is a hospitable vehicle for a selection of insects, including weevils, the saw-toothed grain beetle, and the quaintly named, but no less undesirable, confused flour beetle (figure 5.3). Insecticides, approved for use on the basis of health and safety legislation, have an important role. Like anything else accumulating on the surface of barley, they are washed off during the steeping operation and so don't get into the malt used for brewing.

FIGURE 5.3 Confused flour beetle

Although successive generations of barley varieties tend to have increased resistance to fungal infection, there is still dependence in certain growth regions on the application of systemic fungicides in the field to prevent the development of diseases such as mildew, eyespot, and take-all. These fungicides, by keeping the barley plant free from disease, help it to produce grain that is well filled and in really good condition for malting. Everybody benefits: the farmer, because he enjoys a high crop yield; the Maltster, because she has good viable, healthy, and fragrant barley to malt; the Brewer, because the malt is uniformly of excellent quality and will "behave well" in the brewery, producing excellent beer in good yield, free from materials originating in any infectious agent on grain; and the consumer, because she will be purchasing a quality product with no defects that might be traced to the barley, for example a flavor problem or gushing (see chapter 4).

There is, naturally, considerable emotion introduced when one talks about things like pesticide usage. The "green" movement is burgeoning, so that we have now reached an era when it is possible to buy "organic" beer, produced from raw materials and by means of processes rigorously screened for the use of agents such as those employed to prevent grain infection and infestation. I well recall in the eighties suggesting in a meeting within Bass that we might develop an organic beer, something just done by the old-fashioned and curiously secretive Sam Smith's. My boss's retort was quick and pointed: "Don't be stupid. All of our beer is excellent, safe and wholesome. If you put a trendy 'organic' tag on one brand, would that not make all of our other beers 'non-organic' and presumably inferior? Which they are not."

I do not feel it appropriate here to go into political mode. What I would say is that in pre-pesticide days there were instances of madness induced by infected grain (ergotism or "St. Anthony's Fire") as well as much waste from unusable crops.

The most recent example of a problem invoked by excessive concern about pesticides was a fusarium "scare" in the nineties in North America. Environmentalist pressures led to the outlawing of straw-burning, which had been the common practice to clear out residual plant material after harvest. An aspect of the fire that could have been predicted but was not taken into account was to kill off fusarium. But now that the straw was being plowed back into the fields, this destruction was not happening. The result was fusarium infection in the next year's harvest. Fusarium produces a protein that causes beer to gush, but much more worrying is its production of something called vomitoxin, whose name captures only part of the devastating impact it can have on the body. Of course the alternative strategy in the absence of burning is to use fungicide to kill off the fusarium. Or to use fusarium-resistant barleys—perhaps developed using gene technology. Fire and smoke, pesticide, GMO…or a risk of vomitoxin. Not easy decisions, and not ones to be taken solely on the basis of emotion.

Only pesticides and fungicides that have been rigorously assessed by legislative authorities and subsequently approved should be used. They will have received extensive screening on a health and safety basis, and it will have been demonstrated that they have no damaging effect on the barley, the process, or the product.

In the United States, regulations on the use of pesticides occur at both federal (Environmental Protection Agency, EPA) and state levels, and furthermore bodies such as the Food and Drug Administration and the Occupational Safety and Health Administration have a say in what may and what may not be done.

STEEPING

The purpose of this stage is to increase the moisture content of the grain from 11%–12% to 43%–46% within two days (figure 5.4). Kernels will not germinate if their moisture content is below 32%. A typical steeping regime will comprise an initial water stage for 6 to 16 hours, to raise the moisture content to 33%–37%. An air rest for 12 to 24 hours follows, during which air will be sucked downward through the grain bed to disturb films of moisture on the grain, expose the embryos to oxygen, and remove carbon dioxide produced by respiration, all of which is designed to prevent the embryo from being "suffocated." This will be followed by a second immersion of 10 to 20 hours, which will bring the moisture to the required level.

Water enters the grain through the micropyle, the small opening at the embryo end of the grain. The waxy pericarp/testa layers of the grain prevent access of water at any other point, unless these tissues have been deliberately damaged (see below).

FIGURE 5.4 Steeping (courtesy of Assured UK Malt)

There are no hard-and-fast rules for steeping regimes; they are determined on a barley-by-barley basis by small-scale trials. It must be realized, too, that a barley changes its properties over time. It increases in so-called vigor (the speed of growth essential for malting), which is reflected in enhanced capability for synthesizing enzymes and, therefore, rate of modification of the endosperm. A barley, then, will need to be processed differently in the malting as the year goes on. In some locations, barley is graded before steeping according to size, because different-sized barleys take up moisture at different rates.

Steeping vessels are normally fabricated from stainless steel and most recently have comprised flat-bottomed ventilated vessels capable of holding as much as 250 tons of barley. The steep water (or "liquor," as it is referred to in some parts of the world) is either from a well, in which case it is likely to have a relatively constant temperature within the range 10°C–16°C (50°F–61°F), or from a city water supply, in which case there may be a requirement for temperature control facilities.

The aim of steeping is to achieve a homogeneous distribution of water across the entire bed of grain. The first steep washes a large amount of material off the barley, including dust and leached tannins from the husk. It goes to drain without reuse, and leads to a significant effluent charge.

A range of process aids has been used from time to time to promote the malting operation, and generally they have been introduced either in a steep or on transfer from steeping to germination. A few Maltsters still employ potassium bromate to suppress the growth of rootlets within the embryo, for such growth is wasteful and can also cause matting, which leads to handling problems. Bromate also suppresses proteolysis. The use of bromate is not permitted in the United States.

More frequent worldwide is the use of gibberellic acid (GA), obtained from fermentations with the fungus *Gibberella fujikuroi*. GA is added to supplement the native gibberellins, which are the main hormones of the grain. Although some users of malt prohibit its use and it rarely (if ever) is used in the United States, GA can successfully accelerate the malting process. It tends to be sprayed onto grain as it passes from the last steep on its way to the germination vessel. Some maltsters couple the use of GA with a scarification (abrasion) process, whereby the end of the corn furthermost from the embryo is abraded. This enables water and GA to enter the distal end of the grain, triggering enzyme synthesis and modification in the region that is normally the last part to be degraded. Because these events are also being promoted "naturally" by the embryo, the resultant effect is called two-way modification. It is an opportunity to accelerate the germination process and to deal with barleys that are more difficult to modify.

GERMINATION

The aim of germination is to develop the enzymes capable of hydrolyzing the cell walls, the protein, and the starch and to ensure that these act to soften the endosperm by removing the cell walls and about half of the protein, while leaving the bulk of the starch behind.

Traditionally, steeped barley was spread out to a depth of up to 10 cm (4 inches) on the floors of long, low buildings and germinated for periods of up to ten days, with men using rakes either to thin out the grain ("the piece") or pile it up, depending on whether the batch needed its temperature lowered or raised: the aim was to maintain it at 13°C–16°C (55°F–61°F).

Very few floor maltings survive, because of their labor intensity, although of course there are those who fervently believe that it's the only way to make decent malt. A range of designs of pneumatic (mechanical) germination plant is now used. The newest germination vessels are circular, of steel or concrete, with capacities of as much as 500 tons, and they are microprocessor-controlled (figure 5.5). They may incorporate vertical turners located on radial rotating booms, but just as frequently it is the floor itself that rotates, against a fixed

FIGURE 5.5 A modern germination vessel (courtesy of Stan Sole)

boom. Modern malting plants are arranged in a tower format, with vessels vertically stacked, steeping tanks uppermost.

Germination in a pneumatic plant is generally at 16°C–20°C (61°F–68°F). In this process, some 4% of the dry weight of the grain is consumed to support the growth of embryonic tissues, and much heat is produced. To dissipate this heat demands the use of large amounts of attemperated air, whose oxygen is needed by the embryo for respiration, with the carbon dioxide produced being flushed away by the air flow.

Take a walk through a malting plant with an experienced maltster and you will see him grab a handful of germinating grain and spread it on the palm of one hand, glance at it, and then rub a few corns between the thumb and first finger. If the whole endosperm is readily squeezed out and if the shoot initials (the acrospire) are about three-quarters the length of the grain, then the "green malt" is ready for kilning.

KILNING

Kilning comprises the drying of malt to such a low level of moisture that it is stabilized, with germination arrested and enzymatic digestion halted. The enzymes of the malt, though, must not be destroyed: it is always important that the starch-degrading enzymes survive into malt, because the brewer needs those to generate fermentable sugars in the mash. Often it is important that the cell wall degrading enzymes survive, too, because they may not have completed their job in the malthouse—and they may be needed to deal with polysaccharides present in unmalted adjuncts that the Brewer may use in mashing.

There is a great variety of kiln designs, but most modern kilns feature deep beds of malt. They are most frequently circular in cross section and are likely to be made from corrosion-resistant steel. They have a source of heat for warming incoming air and a fan to drive or pull the air through the bed, together with the necessary loading and stripping systems. The grain is supported on a wedge-wire floor that permits air to pass through the bed, which is likely to be up to 1.2 meters (4 feet) deep.

Of course, kilning is an extremely energy-intensive operation, so modern kilns incorporate energy conservation systems such as glass tube air-to-air heat exchangers.

Newer kilns also use "indirect firing," in which the products of fuel combustion don't pass through the grain bed but are sent to exhaust, the air being warmed through a heater battery containing water as the conducting medium. Indirect firing arose because of concerns with the role of oxides of nitrogen present in kiln gases that might have promoted the formation of nitrosamines in malt. Nitrosamine levels have not been a problem in malt for many years.

Lower-temperature kilning regimes give malts of lighter color and tend to be employed in the production of malts destined for lager-style beers. Higher temperatures, apart from giving darker malts, also lead to a wholly different flavor spectrum. Lager malts give beers that are relatively rich in sulfur-containing compounds. Ale malts have more roasted, nutty characters. For both lager and ale malts, kilning is sufficient to eliminate the unpleasant raw, grassy, and beany characters associated with green malt.

When kilning is complete, the heat is switched off and the grain allowed to cool before it is stripped from the kiln in a stream of air at ambient temperatures. On its way to steel or concrete hopper-bottomed storage silos, the malt is "dressed," which involves mechanical removal of dried rootlets (referred to as "culms," which go to animal feed), aspiration of dust, sifting out of loose husk and incomplete kernels, and elimination of any large contaminants.

TYPES OF MALT

Some malts are produced not for their enzyme content but rather for use by the Brewer in relatively small quantities as a source of extra color and distinct types of flavor (figure 5.6). They may also be useful sources of natural antioxidant materials. There is much interest in these products because of the opportunities they present for brewing different styles of beer. Table 5.2 describes some of these malts, which are produced in small drum kilns equipped with water sprays, for obvious reasons! Those specialty malts produced with the least extra heating (e.g., Carapils and crystal malt) can be used to introduce relatively sweet, toffee-like characters. Those produced with intense heating (e.g., black malt) deliver potent burnt and smoky notes.

WHAT THE BREWER LOOKS FOR IN A MALT

As the years have unfolded and the links have progressively been established between malt composition, the behavior of a malt in the brewery, and the

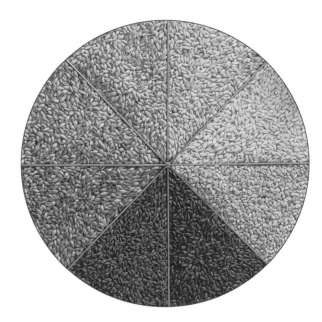

FIGURE 5.6 Types of malt. Clockwise, starting with the darkest colored malt, they are: roasted barley (unmalted); chocolate malt; specially processed malt (kilned and roasted); dark caramel malt; light caramel malt; Munich malt; 2-row pale malt; wheat malt (courtesy of Briess Malt & Ingredients Company)

TABLE 5.2. TYPES OF MALT

Product	Details	Purpose/comments
Pilsner malt	Well-modified malt, gentle kilning regime not rising above ca. 85°C (185°F)	Mainstream malt for pale lager beers
Vienna malt	Similar to Pilsner malt but higher protein, more modification, final kilning temperature ca. 90°C (194°F)	Mainstream malt for darker lagers
Munich malts	From higher-protein barleys (e.g., 11.5% protein), prolonged germination, low-temperature (e.g., 35°C, 96°F) onset to kilning to allow stewing (ongoing modification), then rising temperature regime to curing at over 100°C (212°F)	For darker lager beers
Pale malt	Relatively low N (e.g., <10% protein), well-modified, kilning starting at ca. 60°C (140°F) and rising to a final curing temperature ca. 105°C (221°F)	Mainstream malt for pale ales
Chit malt	Very short germination time and lightly kilned	Permissible as adjuncts in countries such as Germany with restrictions such as Reinheitsgebot
Green and lightly kilned malts	No or restricted kilning after substantial germination	Alternate to exogenous enzymes
Diastatic malts	High-protein barley (especially 6-row), steeped to high moisture content, long cool germination, gibberellic acid if permitted, very light kilning	High enzyme potential for use in mashing with high levels of adjuncts
Smoked malts	Kilning over peat	For beers with smoky character, e.g., rauchbier

Malt	Process	Purpose
Wheat malts	Germinated wheat, usually somewhat under modified, lightly kilned (e.g., < 40°C, 104°F)	For wheat beers
Rye malts		For specialty beers
Oat malts		For specialty beers, including stouts
Sorghum malts	Steeps may incorporate antimicrobials such as caustic; warm germination (25°C; 77°F)	For sorghum beers; malted millet may also be used as a richer source of enzymes
Carapils (a caramel malt)	The surface moisture is dried off at 50°C (122°F) before stewing over 40 minutes with the temperature increased to 100°C (212°F), followed by curing at 100°C–120°C (248°F) for less than 1 hour	To afford color and malty and sweet characters to lighter beers
Amber malt	Pale malt is heated in an increasing temperature regime over the range 49°C to 170°C (306°F)	To afford bread crust, nutty characters to beer and color
Crystal malt	As for Cara Pils, but first curing is at 135°C (275°F) for less than 2 hours	To afford toffee, caramel characters to beer, plus color
Chocolate malt	Lager malt is roasted, by taking temperature from 75 to 150°C (302°F) over 1 hour, before allowing temperature to rise to 220°C (428°F)	To afford chocolate, roast, coffee, burnt, bitter characters to beer, plus color
Black malt	Similar to chocolate malt, but the roasting is even more intense	To afford harsh, astringent, roast, burnt notes to beer, plus color

Modified from M. J. Lewis and C. W. Bamforth, *Essays in Brewing Science*, Springer (2006) quality of the finished beer, more and more demands have been placed on the Maltster by the Brewer. There has been a tendency, still prevalent among the traditionalists, for the Brewer to blame the poor old Maltster whenever things go wrong. The trend, too, has been for the Brewer to place more and more specifications on the malt, in many cases rather unfairly because the demands are often contradictory.

Many Brewers, applying quality assurance principles, will look to the Maltster to provide documentation with the malt shipment that details all the required analyses on a batch of malt, certifying that the malt meets the required tolerances. The Brewer will spot-check occasional batches. Heaven protect the transgressing Maltster! From time to time, too, the Brewer will audit the Maltster (as, indeed, he will audit most of his raw material suppliers). I don't believe that there is a Brewer today who applies the sliced bread test for auditing malting hygiene; time was when the Brewer would visit the malting, wipe the inside of a vessel with a piece of bread, and invite the Maltster to eat it for breakfast. Crude certainly, but a powerful incentive for the Maltster to keep the plant spotless!

Thankfully, most of the analytical methodology applied to assess malt quality is somewhat more sophisticated than this.

This chapter has shown how barley and water have come together meaningfully in the production of the soul of beer. The demand for water has been great, and so it is, too, in the brewery. So what do we expect from that water? We will find out in the next chapter.

WATER

AND GENUINE TERROIR

An old boss of mine once described beer as "slightly contaminated water." This of course is an offensive point of view, but the reality is that few beers comprise less than 90% water. It is therefore the brewer's principal ingredient, and one upon which she should lavish abundant attention.

We must appreciate, too, that much more water is used in a brewing operation than ends up in the beer. Treated water is needed for brewing and processing, but also for dilution and cleaning. Softened water is required for the washing of bottles and for the pasteurizer. Demineralized water is demanded by the boiler, while plenty of untreated water is sloshed around to clean floors and so on. Environmentally aware and cost-conscious Brewers lavish attention on conservation endeavors. It is generally held that a brewer is faring well if using six times more water than ends up in the beer. The best brewers are approaching half that usage. The cost of water and ensuing wastewater treatment or disposal is similar to that of energy, an expense item that tends to more readily garner attention.

It is usually relevant to discuss water supply and wastewater treatment together, for the simple reason that here, unlike with malt or hops, we have a raw material that is recyclable. Indeed the extent to which wastewater is generated is under increasing pressure through the relentless march of treatment charges, but also the never lessening dictates of legislation. Water disposal in rural areas is quite straightforward if the ground is pervious, but is seldom straightforward in urban districts, especially those featuring industry.

Wastewater passes either to a treatment works or, depending on the location, well out to sea, where the combination of high salt level, wave action, and microflora destroy even the biggest lumps. Treatment works comprise solid filtration, chemical precipitation, and bacterial action, promoted by ample aeration. A number of Brewers nowadays have their own aerobic or anaerobic waste treatment plants, thereby delivering back to the water supplier a greatly reduced effluent load.

Once I was in a cab in a major city on the western edge of Europe, in the midst of an oil crisis. The taxi driver was deadly serious in pronouncing his amazement at the kerfuffle, considering that petrol simply "comes up from underground into the pumps." While it can be exactly that scenario in the case of water, notably for us in the guise of brewers' own wells, as often as not the provision of water is somewhat more complicated than this. The trend in brewing is toward acquiring water from a municipal supply, it being frequently more economic in the light of increasing legislation to offload the task of providing good wholesome water to the local authorities.

The Earth's surface is 71% water. It is in constant flux, with evaporation from the oceans and precipitation as rain or snow. Some of the precipitation runs off on the surface into rivers, some reevaporates, and the remainder percolates underground. Yet more water rises from the subterranean depths and is expelled as "juvenile water" from minerals via physical and chemical reactions. This water frequently emerges at the surface as hot or cold springs and flows away as rivers.

There are invariably dissolved substances in all natural water. These materials define that water and dictate its provenance, but may be less than ideal when the water is used for manufacturing purposes, such as the production of beer. There are minerals in spring water, salt in sea water, and stinky sulfurous gases in spa water. Rainwater contains dissolved oxygen and carbon dioxide from the air, as well as atmospheric pollutants such as oxides of nitrogen. The groundwater at or near the surface will frequently contain organic products from decay reactions, together with many living organisms. It is said to be "stagnant."

Seawater can contain up to 3% sodium chloride. The water in estuaries and coastal marshes is said to be "brackish": it contains less salt but enough to be tasteable. According to the World Commission for Water in the 21st Century (2000), only 2.5% of the world's water is not salty, and two-thirds of that is frozen in glaciers and ice caps. Of the amount that's left, 20% is found in excessively remote areas, and of the remainder, 75% comes at the wrong time and in the wrong place as monsoons and floods. Thus, in toto, just a tiny fraction of 1% of the total planetary water is actually "easily" available.

Water flowing off igneous rocks (the primitive rocks from the cooling of the original magmas and lava from the developing molten earth) contains little dissolved material, quite unlike water from sedimentary rocks (those arising from mineral debris transported from igneous rocks). In the latter instance, bicarbonates and sulfates of calcium and magnesium confer soap-destroying activity and the tendency to produce scale in pipes and kettles. This, of course, is the so-called hardness, of which more momentarily.

Some natural water, especially that from springs and deep wells, is potable without any additional treatment. However, when water is in bulk supply and destined for storage, it is invariably purified. Water is delivered from river, lake, or well via pipes or aqueducts to "impounding" reservoirs, which are essentially buffers. These are open and may have a diversity of uses, such as recreational. They may be stocked with fish. In these reservoirs there is a settling of sand and other particles and a lowering of undesirable microflora.

Water is drawn off from the impounding reservoirs as needed and delivered through pipes to the waterworks. Here the processes comprise sedimentation, coagulation, filtration, and disinfection, en route to a covered "service" reservoir, which ideally is located at a sufficient altitude to provide a head of pressure that will drive subsequent flow. (Here in the States, we are constantly reminded of this law of physics through the abundance of water towers.)

The next stage is clarification, involving either deep bed filtration through sand or coagulation with an agent such as aluminum sulfate. Chances are that the water will still be cloudy, so it will be filtered through a mix of clean sand, gravel, and pebbles.

Finally, the water will be disinfected using traces of chlorine (such as 1 ppm for a contact time of 1–2 hours) or ozone, ultraviolet, or lime.

In the United States, water must satisfy the National Primary Drinking Water Regulations (http://www.epa.gov/safewater/contaminants/index.html).

TEMPORARY OR PERMANENT HARDNESS

The hardness of water can be assessed by seeing how easy it is to generate a lather using soap. "Soft" water, which contains very low levels of minerals, lathers readily. But "hard" water doesn't. If the hardness is eliminated by boiling, it is said to be "temporary hardness": it is due to carbonates and bicarbonates of calcium and magnesium. If, however, boiling has no impact, then it is permanent hardness, and it is due to sulfates of calcium and magnesium. Another way to tell whether your local water is hard is to check for scale in a water heater or kettle. The scale is due to calcium and magnesium-rich deposits.

Water needs to be softened if scale is to be avoided in boilers, and, of course, some people believe that softening needs to be performed to a greater or lesser extent for brewing, depending on the style of beer being produced. Softening can be achieved by various chemical treatments or by using agents called ion-exchange resins.

The fastidious brewer is ever more concerned with getting the very last traces of insoluble material as well as undesirable chemical materials out of the water, which in reality means tighter and tighter regimes of filtration and adsorption. Concerns with incoming water may include levels of nitrates, which may be converted to nitrosamines by bacteria in the brewhouse, and the presence of pesticides, foul-tasting chlorinated hydrocarbons, and protozoa such as *Cryptosporidium* and *Giardia*.

The conscientious Brewer should be mindful of the local risks to the water supply, whether this is their own wells or a municipal supply. There is no lessening of these threats, with increasing urbanization, industrialization, and intensification of agriculture.

I fear that there are some Brewers who are rather too blasé about their water supply. True, its analysis and the greater part of the burden in ensuring its wholesomeness lie with the supplier. Yet this should never excuse the Brewer from caring passionately about it, just as much as he or she scrutinizes the malt, the hops, and the yeast.

At the very least it should be taste, taste, taste—not only the incoming water, but the water after every process stage that it has been put through: charcoal filtration, ion exchange, deaeration, even storage.

Quite apart from the obvious requirements, such as an absence of taints and of hazardous components and an adherence to all requirements as a potable supply (to satisfy all of which a Brewer may treat all water by procedures such as charcoal filtration and ultrafiltration), the water must have the correct balance of ions. Traditionally, breweries producing top-fermented ales were established in areas, such as Burton-on-Trent in England, where the level of calcium in water is relatively high (about 350 ppm). This compares with a calcium level of less than 10 ppm in Plzeň, a place famed for its bottom-fermented lagers. In many places in the world the salt composition of the water (often brewers call it "liquor") is adjusted to match that first used by the monks in Burton in the year 1295. This adjustment process is called Burtonization. Sometimes the brewer will simply add the appropriate blend of salts to achieve this specification. To match Pilsen-type water it is necessary to remove existing dissolved ions by deionization.

Calcium in brewing water plays several roles. First of all, it promotes the action of α-amylase. It reacts with phosphate in the malt to lower the pH to the appropriate level for brewing. Also, it precipitates another natural component of malt, oxalic acid, which otherwise would come through into the beer and cause problems such as the blocking of dispense pipes ("beer stone"). Calcium also promotes the flocculation of yeast.

Many brewers (as we have seen) worry about two other ions contributed by water, chloride and sulfate.

So we have learned about the soul of beer (malt) and about water, which might be beer's body. It is time to turn to the spice of beer.

THE WICKED AND PERNICIOUS WEED

HOPS

When many people consider beer, they automatically think of hops. Lots of people are misguided and believe that it is hops alone that are the basic raw materials for making beer, with all of the alcohol and flavor flowing from them. Of course it is the starch in malted barley and adjuncts that serves as the fermentation feedstock that yeast uses to make alcohol. Hops are simply a flavoring material, albeit one that makes other key impacts on beer and brewing. However, as we shall see (and have seen, in chapter 4), the chemistry of hops and hopping is anything but simple.

THE HISTORY OF HOPS

Not until the twelfth century do we find mention of the use of hops in making beer. Indeed beers at that time were flavored with all manner of herbs, including rosemary, yarrow, coriander, and bog myrtle, that were added in mixtures known as gruit. There has even been reference to the use in beer of caraway, pepper, pine roots, spruce, potato leaves, and tobacco. Some added strychnine as a preservative component.

As we saw in chapter 1, it was in the thirteenth century that the hop started to threaten gruit as a flavoring for beers in Germany. "Threaten" is a word I use advisedly, for growers in all countries of the hitherto traditional flavorings fought vigorously against the introduction of hops. The plant was banned from use in brewing in Norwich, England, in 1471 and hops were condemned by the English as being a "wicked and pernicious weed."

Medieval adherents of ale (which was a term then restricted to unhopped beer) will also have rebelled, if not necessarily the brewers. The ales that drinkers were used to were strong and sweet—and deliberately so, for high concentrations of sugar and alcohol suppress the growth of the microorganisms that can ruin beer. Hops, though, have strong antiseptic properties. Using them in beer will have enabled brewers to "thin out" the drink and make it weaker. The first hopped beer wasn't seen in England until the late fourteenth century.

Hops started to be grown in South East England in 1524, 100 years before they were first cultivated in North America. Just as the Yakima Valley in Washington State is famed for its hops, so too is Kent, "The Garden of England."

The tirade against hops was relentless. Andrew Boorde wrote in 1542:

> Bere is made of malte, of hoppes, and water; it is a naturall drynke for a Dutche man, and nowe of lete days it is much used in England to the detryment of many Englysshe men; specially it kylleth them the which be troubled with the colyke; and the stone, and the strangulion; for the drynke is a colde drynke, yet it doth make a man fat, and doth inflate the bely, as it doth appere by the Dutche mens faces & belyes. If the bere be well served, and be fyned, & not new, it doth qualify heat of the liver.

Not a particularly supportive reference (and humble felicitations to my friends at Heineken). Happily, a somewhat different view of hops and their contribution to beer quality now exists, and these days very little malt-based beer is devoid of hops. However, the manner by which the unique bitterness and aroma of the plant is introduced into the beverage is often very different from that which was practiced six centuries ago.

A SOLITARY OUTLET

The hop (figure 7.1) is remarkable among agricultural crops in that essentially its sole outlet is for brewing, apart from a somewhat limited market for its oils in aromatherapy and for whole cones in hop pillows (it is said that if you sleep on such a pillow you will not only sleep well but also dream about your true love). The Romans used hops as a spring vegetable, though quite how popular it was when used in that way is lost in the fogs of time.

Although hopping accounts for much less than 1% of the price of a pint of beer, it has a disproportionate effect on product quality, and, accordingly, much attention has been lavished on hop and its chemistry.

Hops are grown in all temperate regions of the world (table 7.1). Over 100,000 tons are grown each year, approximately one-third of those in Germany, with the United States being the next largest producer, the bulk of cultivation being in three states: Washington, Oregon, and Idaho. Hops are grown in the Southern Hemisphere as well as the North, with a significant crop in Australia, notably Tasmania.

FIGURE 7.1 Hop cones (courtesy of Trevor Roberts, S. S. Steiner)

TABLE 7.1. PRODUCTION OF HOPS (1998)

Country	Hop production (thousand hectares)
Germany	21
United States	18
China	7
Czech Republic and Slovakia	6
Poland	3
Slovenia	3
England	2
Russia and Ukraine	2
Australia	1
France	1
Spain	1

Values are rounded to nearest whole number

There are two separate species of hops: *Humulus lupulus* and *Humulus japonicus.* The Romans called hops *Lupus salictarius, Lupus* meaning "wolf" and the hop being likened to a wolf in an osier bed on account of the manner by which it grew wild amid willows. *H. japonicus* contains no resin and is merely ornamental. Hops for brewing are within *H. lupulus,* which is rich in resins and oils, the former being the source of bitterness, the latter the source of aroma.

The hop genus (*Humulus*) is within the family *Cannabinaceae*, and a close relative of the hop is *Cannabis sativa*, Indian hemp, better known as marijuana or hashish. A key point of distinction is in their respective resins: those from hops make beer bitter; those from marijuana have hallucinogenic effects, or so I am led to believe. I believe there may even be folks who smoke hops, though I don't know any of them myself.

CULTIVATION

Hops are hardy climbing herbaceous perennial plants. They are grown in yards using characteristic string frameworks to support them (figures 7.2 and 7.3). Their rootstock remains in the ground from year to year and is spaced in an appropriate fashion for effective horticultural procedures (e.g., spraying by tractors passing between rows). In recent years, so-called dwarf varieties have been bred, which retain the bittering and aroma potential of "traditional" hops but grow to a shorter height (6–8 feet as opposed to twice as big). As a result, they are much easier to harvest and there is less wastage of pesticide during spraying. Dwarf hop gardens are also much cheaper to establish.

FIGURE 7.2 Hop trellis work (courtesy of Stéphane Meulemans, Yakima Chief)

Hops are susceptible to a wide range of diseases and pests. The most serious diseases are *Verticillium* wilt, mildew, and mold, with the damson-hop aphid an especially unwelcome visitor. Varieties differ in their susceptibility to infestation and have been progressively selected on this basis. Nonetheless, it is frequently necessary to apply pesticides, which are always stringently evaluated for their influence on hop quality, for any effect they may have on the brewing process, and, of course, for their safety. A Brewer will not use hops or hop preparations (or, indeed, any other raw material) unless absolutely convinced that they will be entirely hazard-free for process, product, and consumer.

Some hop-growing regions present more of a problem in terms of diseases and pests than do others. For instance, whereas mildew has regularly been of concern in Europe, it had virtually been unheard-of as a problem in the United States after it was first observed there in 1909 in hop gardens on the East Coast, a finding that precipitated the shift of the hop-growing business to the opposite coast. Unheard-of, that is, until 1997, when more than half of the hop crop in the Yakima Valley succumbed to powdery mildew.

FIGURE 7.3 Hops in growth (courtesy of Stéphane Meulemans, Yakima Chief)

The components of the hop required by the brewer—the resins and the oils—are located in the cones of female plants. More particularly, they are found in the lupulin glands (figure 7.4).

Internationally, there are differences in preference for seeded as opposed to unseeded hops. In the United Kingdom, male plants may be included alongside female plants, leading to fertilization and seed levels of up to 25%. Hops are perennial, however, and can be propagated from cuttings, so, unfortunately for the male of the species, his services can be readily dispensed with. Indeed, in the rest of Europe and the United States (with some exceptions in Oregon), there is no planting of male hops, and the hops supplied for brewing are seedless. On a weight-by-weight basis, the content of resin and oil is greater in seedless hops, but horticultural yield is lower. It is believed by some that seedless hops make for easier downstream processing of beer.

A typical grower's year in the hop-growing district of Yakima in the state of Washington will commence in March with some shallow plowing to lower the weed count and to mulch into the ground residual leaves and vines from the previous crop, as well as some fertilizer. In the following month, the wirework will be established on 3-m wooden poles at a spacing of 2 m × 2 m, prior to training new shoots from the rootstock onto the strings. In June, plowing will be undertaken to control weeds, while spraying will be employed in July and August to control pests. Harvest will commence in mid-August and last for a month.

Hops are harvested within a similar time span in the Kent and Hereford-and-Worcester hop gardens of England, but through the month of September. Traditionally, this has been a most labor-intensive operation, demanding short-term labor bussed in from the city, but now machine picking is universally employed.

Drying of hops in kilns is according to similar principles to that for barley and malt (see chapter 5). Using temperature regimes of between 55°C and

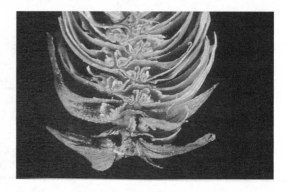

FIGURE 7.4 Cross section of a hop cone showing lupulin glands (courtesy of Trevor Roberts, S. S. Steiner)

65°C (131°F–149°F), the moisture content will be reduced from 75% to 9%. Traditionally, in the United Kingdom hops are delivered to the Brewer compressed in jute (now polypropylene) sacks about seven feet deep. These are called "hop pockets" (not to be confused with "hot pockets"!), each holding 75 kilograms. In the United States, hops are packed into bales weighing 200 pounds. The hops must be stored cold and in an air-tight condition, to counter the degradation of the bitter substances and the attendant development of nasty cheesy characteristics. As we shall see, though, the use of hop cones without some form of modification is rare these days.

HOP ANALYSIS

As for all raw materials used in the brewing process, specifications are applied to hops that must be met if a transaction is to be conducted between the hop merchant and the Brewer. It is fair to say that the analysis of hops remains somewhat more primitive than that of other brewery ingredients. Many of the assessment criteria for hop quality depend on noninstrumental judgment by experts. First the sample of hops will be inspected visually for signs of deterioration, infestation, or weathering. Then a sample will be rubbed between the palms of the hands before sniffing the contents; the assessor is looking for any smells associated with deterioration and, just as importantly, is determining whether the "nose" is consistent with that which is expected from the variety in question. The buyer is not looking for a character that will manifest itself in the finished beer, for the aroma of hops and that of hoppiness in beer seldom bear a simple relationship. Rather, he is looking for certain varietal characteristics in the aroma to confirm that the hop is what it is purported to be. The oil component of hops ranges from just 0.03% to 3% of the weight of a hop. Seedless hops tend to contain more essential oil. The oils are produced in the hop late in ripening, after the majority of the resin has been laid down, which highlights the need for harvesting of the hops at the appropriate time.

The content of resins in hops (and therefore the potential bitterness) can be quantified by titration or by measuring the amount of ultraviolet light trapped by solutions made from the hops. For those hops being purchased expressly for the provision of bitterness, the cost of the hop is assessed on the resin content.

TYPES OF HOPS

There is an increasing tendency to classify hops into two categories, aroma hops and bittering hops. In reality, they are merely variations on a theme.

All hops are capable of providing both bitterness and aroma. Some hops, however, such as the Czech variety Saaz, have a relatively high ratio of oil to resin, and the character of the oil component is particularly prized. Such varieties command higher prices and are known as *aroma varieties*. They will seldom be used as the sole source of bitterness and aroma in a beer: a cheaper, higher-resin hop (a *bittering variety*) will be used to provide the bulk of the bitterness, with the prized aroma variety added late in the boil for the contribution of its own unique blend of oils. Those Brewers requiring hops solely as a source of bitterness may well opt for a cheaper variety, ensuring its use early in the kettle boil so that the provision of bitterness is maximized and unwanted aroma is driven off.

In just the same way that there are many varieties of malting barley and fierce loyalty is shown by Brewers to one or a very few of these varieties, the selection of hops is a serious concern for the Brewer. In some countries one variety prevails, as is the case in Australia, where Pride of Ringwood has held center stage for many years. This situation can be compared with that in the United States, where a bittering variety, Cluster, has been in use since 1800, joined these days by the likes of Eroica, Galena, and Nugget, whereas aroma varieties like Mount Hood have emerged in the last ten years. Cascade, with its characteristic grapefruit notes, is highly prized—for example, in the excellent Sierra Nevada Pale Ale. The U.S. market is an interesting mix of modern and traditional, for prized aroma varieties like the English Fuggles and German Tettnang have been in use for over a century. Some of the varieties are classified as "dual purpose," being considered useful both for bitterness and high-quality aroma potential. These include Chinook and the German variety Perle. Some consider Cluster to be in this category.

The history of Fuggles (first discovered in 1861 as a seedling amid crumbs from a hop picker's lunch) is a good example of the pressures that drive the hop market. It was introduced commercially in 1875 in Kent, and half a century ago accounted for 75% of the English hop crop. Its problem is an acute susceptibility to *Verticillium* wilt. Breeding programs have delivered just one variant of Fuggles (Progress) that shows a sufficiently improved resistance. Accordingly, programs have focused on seeking in other varieties the necessary blend of quality and disease-resistance characteristics. As yet, no single variety displays a comprehensive resistance to all diseases while simultaneously displaying high bitterness potential and good aroma.

The most famed hop-growing region is the Hallertau, to the north of Munich in Germany, where a hop garden was first reported in the year 736. The western part of the Czech Republic, a region known as Bohemia, is also feted for its hops.

WHAT HAPPENS TO THE HOPS IN THE BREWERY

When wort is boiled in the kettle (see chapter 8), the resins are chemically rearranged to soluble and bitter forms in a process referred to as isomerization. The process is not particularly efficient, with perhaps no more than 50% of the resins being altered in the boil and less than 25% of the original bittering potential surviving into the beer.

The oil is a complex mixture of at least 300 compounds. Nobody can yet claim to have established a clear relationship between the chemical composition of the essential oils and the unique aroma characteristics that they deliver. It is most likely that "late hop character" (the aroma associated with lagers from mainland Europe, which is introduced by adding a proportion of the hops late in the boil) is due to the synergistic action of several oil components, perhaps modified by the action of yeast in fermentation. "Dry hop character" (a feature associated with traditional English cask ales, afforded by adding a handful of whole hop cones to the vessel) is no less complicated. To a greater or lesser extent, the individual essential oil components are lost from wort during boiling. The delivery of a given hop character, then, depends on the skill of the brewster in adding the hops at exactly the right time to ensure survival of the right mix of oils that imparts a given character to her product. No instrumental method is available as yet to assist in this process.

HOP PREPARATIONS

The use of whole cone hops is comparatively rare nowadays (figure 7.5). The most common procedure for hopping is to add hops that have been hammer-milled and then compressed into pellets. In this form they are more stable and more efficiently utilized, and they do not present the brewer with the problem of separating out the vegetative parts of the hop plant.

Nevertheless, because of the inefficient utilization of the resins during wort boiling, even from pellets, and as a result of vagaries in the introduction of defined hoppy aromas into beers, a wide selection of hop preparations have reached the marketplace. We can actually trace proposals for making hop extracts back to 1820, when lupulin glands were extracted with lime, alcohol, and ether. Nowadays, extracts are mostly based on the prior extraction of hops with liquid carbon dioxide.

It was first shown over thirty years ago that the resins and oils of hops could be extracted using as a solvent carbon dioxide that has been liquefied at high pressure and low temperature. The resultant extracts can be fractionated into resin- and hop oil–rich fractions, with the resin portion being available

FIGURE 7.5 Hop pellets and hop extract (courtesy of Stéphane Meulemans, Yakima Chief)

as a source of bitterness for addition in place of whole hops or pellets to the kettle and the oil part providing an opportunity for controlled addition of hop character, either by dosing late in the boil for a late hop character or into the finished beer for a dry hop note.

It is possible to carry out the isomerization of the resins in the liquid CO_2 extracts by chemical means or by the use of light. Therefore, it is possible using the resultant "pre-isomerized extracts" to add bitterness directly to the finished beer, which makes for far better utilization of the bitter compounds because the extent of isomerization of resins is greater and because bitter substances are no longer lost by sticking onto yeast cells. A sizable quantity of beer is brewed worldwide in which all of the bitterness is introduced in this way.

Recent years have been marked by an enormous increase in the use of such pre-isomerized extracts after they have been modified by a process known as "reduction." The isomerized resins are susceptible to cleavage by light, resulting in pronounced skunky character in beer. If the resins are reduced, they no longer produce this noxious aroma. For this reason, beers that are destined for packaging in green or clear glass bottles are often produced using these modified bitterness preparations, which have the added advantage of possessing increased foam-stabilizing and antimicrobial properties.

Late hop aroma can be introduced through the use of extracts, too. It has been shown that the essential oils can be split into two fractions, one of which is spicy and the other floral. By adding them to bright beer in different proportions, it is possible to impart different late hop characters, again offering tremendous opportunities for new product development. Here is a mechanism for the brewer to introduce under controlled conditions a range of flavor characteristics to beer and potentially to create a selection of different products by downstream adjustment of a single base beer.

We can see that the extraction of hops to make products such as pre-isomerized extracts, reduced resins, and late hop essences has introduced enormous opportunities for a Brewer. Each of these materials is added as late as possible in the process. Still, though, most of the hopping of beers is carried out in the brewhouse, which is where we turn now.

COOKING AND CHILLING

THE BREWHOUSE

The production of beer can be conveniently split into a "hot side" and a "cold side." The former takes place in the brewhouse, the latter in the fermentation cellar and all points downstream of it. Strictly speaking, *brewing* is what happens in the brewhouse, a process designed to convert malt and any adjunct materials into a liquid called *wort* (rhymes with "Bert"), which will form the feedstock that yeast will convert into alcohol. Traditionally (and still extensively), it is in the brewhouse that the hops are introduced into the process.

CHEMISTRY AND BIOCHEMISTRY IN THE BREWHOUSE

Wort needs to have various features. First, it must contain sugars that the yeast is capable of fermenting into alcohol. These sugars are the energy source that the yeast needs to support its growth. It is not a question of any old sugars: the balance of different types can have a profound effect on the way that yeast behaves and the efficiency with which it converts them into alcohol. Moreover, the type of sugar influences the balance of flavor compounds that the yeast produces, and therefore the taste of the beer.

Secondly, the yeast requires from wort the building blocks that it will use to synthesize its proteins. These are the amino acids, which in turn are produced during malting and mashing by the breakdown of barley proteins. Once more, they must be in balance. The relative proportions of the different amino acids influences yeast behavior, as does the ratio of sugars to amino acids.

The balance of sugars and amino acids is determined by what happens in the brewhouse. It is also within the brewhouse that the brewer establishes the right salt balance in the wort, whether or not the wort will contain the necessary levels of sulfur and other elements that the yeast depends on, and whether the beer will contain the necessary levels of foaming materials. And it is here that a range of undesirable materials is eliminated, including unpleasant flavors and materials that can promote turbidity in beer.

With the development of ingredients such as pre-isomerized hop extracts, it is no longer the case that bitterness is necessarily determined in the brewhouse, but this is still the practice in a great many breweries. Color, too, can be modified downstream, but for as long as the brewhouse is a standard feature of brewery operations, it will have a major impact on all aspects of downstream performance and product quality.

We are about to embark on a simple description of the enzymatic processes that are involved in mashing (and have to some extent already begun in mashing). The Appendix may prove useful now, to those for whom enzymology is a mystery.

THE BREAKDOWN OF STARCH

As we have seen in chapter 5, it is generally important for efficient brewhouse operation that the grain has had its cell walls comprehensively degraded, as well as perhaps half of its protein. It is essential, however, that the bulk of the starch within the endosperm survive malting, for it is this which the brewster will be using as a source of fermentable sugar to "feed" her yeast. The remarkable fact is that this is seldom a problem and that starch by and large does survive the malting process, even though the enzymes needed to disrupt it are plentiful. This tells us that starch is a relatively tough nut to crack. If it is to be broken down in the relatively short time frames available to a brewer (frequently no more than an hour in the mash), then it must first be "gelatinized." When starch granules are heated, the molecules of which they are composed "melt" and the granular structure disaggregates. This melting (or "pasting") occurs at different temperatures depending on the origin of the starch. For barley starch it is usually complete once the temperature gets to 65°C (149°F).

Mashing, then, usually includes a "conversion" stage, typically at 65°C (149°F), for 50–60 minutes. The starch will be gelatinized almost immediately, rendering it accessible to attack by the amylase enzymes that rapidly hydrolyze it.

Rice starch gelatinizes over the range 64°C–78°C (147°F–172°F) and corn starch at 62°C–74°C (144°F–165°F). These cereals, if they are used as adjuncts in the brewery, must therefore be "cooked," and the brewhouse that uses them will have a cereal cooker alongside the mash vessel. Wheat starch gelatinizes at temperatures similar to barley starch, and therefore wheat flour can be used directly in the mash.

Within the starch granules there are two populations of starch: the amylose and the amylopectin. Both of these molecules are polymers of glucose units linked together in chains. They differ in that the molecules of amylose, which typically amount to 25%–30% of the total starch, are linear chains of perhaps 1,800 glucoses, whereas amylopectin molecules comprise many much shorter chains linked through branch points. The significance of this is that they require different enzymes to chop them up.

The major starch-degrading enzyme in malt is α-amylase. It's very similar to the enzyme found in human saliva—indeed, in some societies fermentation of alcoholic beverages begins with the starch being digested by a generous donation of saliva from the "brewer." Thankfully, I am not aware of any beer brewers presently applying this technique! The α-amylase attacks at random in the middle of the amylose and amylopectin, releasing some small sugars but primarily short chain molecules (called dextrins). The next enzyme is the β-amylase, which starts at one end of the dextrin molecules, chopping off two glucoses at a time. (Two glucoses joined together represent the sugar *maltose*, so named because it is the major sugar found in mashed malt.)

With amylose, the combined action of these two amylases leads to a mixture of sugars that is completely fermentable. Such is not the case with amylopectin. Its branch points are not chopped up by either of these amylases, and when β-amylase encounters them, it can't get past them. A third enzyme is needed, one whose role is to hydrolyze the branches. It is called limit dextrinase, and it is only produced late in the germination process. Moreover, it is bound up with other components from malt, which limits its activity. The outcome is that conventionally mashed malt doesn't produce totally fermentable wort, with perhaps 20% of the sugar being tied up in a dextrin form. Most beers worldwide contain residual dextrin for this reason, and the belief is that these dextrins contribute to the body of beer. The so-called diet or light beers, however, contain no residual sugar. The brewer may have added heat-stable enzymes (from microbial sources such as *Aspergillus*, a food-grade organism used, for example, in brewing sake) that are capable of chopping up the branch points in amylopectin. As a result of the combined efforts of the malt-derived and the exogenous enzymes, all of the starch is converted into fermentable sugar. Alternatively, a much more complex mashing regime may be used to maximize the opportunity for the various starch-degrading enzymes to act, perhaps followed by the inclusion of a small charge of an extract of very gently kilned malt in the fermenter so as to introduce more of the enzymes needed for completion of starch degradation.

Some of the low-alcohol beers in the market, containing perhaps 1%–2% alcohol (by volume), are produced using a technique called "high-temperature mashing." If the malt is mashed at a higher than normal temperature, say 72°C

(162°F), the β-amylase is quickly destroyed and far less maltose is produced in the wort. Most of the starch is converted only as far as nonfermentable dextrins, and so the resultant wort contains much less sugar transposable by yeast into alcohol.

THE BREAKDOWN OF CELL WALLS

Most Brewers look to the Maltster to provide them with malt that has had its cell walls comprehensively removed. In practice, most malts have some cell wall material remaining, either intact or partially degraded, and this can cause problems.

The cell walls of barley contain two major polysaccharides: the β-glucans and the pentosans. The former, which account for some 75% of the wall, consist of a straight chain polymer of glucose, just like amylose. The difference is in the way that the glucoses are joined together. This means that the properties of β-glucans and starch are very different, and also that a totally distinct set of enzymes is needed to break down the two materials. Pentosan is also a sugar polymer, but this time the backbone is a chain of sugars that each contain five carbon atoms (hence they are called *pentoses*—"pent-" as in *pentagon*— as opposed to a sugar such as glucose, which contains six carbon atoms and so is called a *hexose*—"hex-" as in *hexagonal*).

These β-glucan and pentosan molecules give very viscous solutions. If they are not broken down in malting or mashing, they will be extracted into wort to cause all manner of problems for the brewer because of this viscosity effect; there will be a slowing down of the rate at which the wort can be separated from the spent grains (see below) and, because these molecules will survive fermentation intact, they will get into beer and greatly reduce rates of beer filtration. Because beer is filtered around 0°C and viscosity increases as temperature is lowered, this is a particular problem. Not only this, but the solubility of the cell wall polymers is reduced as the temperature falls, and if this material survives into beers there is the risk of sediment formation in beers stored in refrigerators. This is a particular problem with stronger beers; because they contain more alcohol, they are likely to have been made from more concentrated worts (in other words, more malt per unit volume). In turn, this greater "malt contribution" will yield higher levels of molecules such as β-glucans and pentosans to the beer. The situation is compounded further by the fact that alcohol itself acts as a precipitant, increasing the likelihood that the β-glucan and pentosan will collect in the bottom of the bottle as fluffy sediment.

The most important enzyme from malt that degrades the more troublesome of the polymers (β-glucan) is β-glucanase. It is produced in ample quantities early on in germination, and, providing it gets distributed through the starchy endosperm in malting, it is capable of removing most of the β-glucan. Most, but not all. The major problem with this enzyme is that it is extremely sensitive to heat. At 65°C (149°F, the temperature used to gelatinize malt starch), this enzyme is totally destroyed in a couple of minutes. For this reason, many brewers commence their mashing operation at a relatively low temperature (say, 45°C–50°C, 113°F–122°F) to enable the β-glucanase to act, and then, after twenty minutes or so, the temperature is ramped up to 65°C (149°F). Alternatively, a heat-resistant food-grade β-glucanase, perhaps from the bacterium *Bacillus subtilis* or the fungus *Penicillium funiculosum,* can be added at the conversion temperature.

THE BREAKDOWN OF PROTEIN

Just as for cell walls, it is the malting operation that is most significant for protein hydrolysis (or proteolysis). The brewster does not want total degradation of protein, for unlike the cell wall polysaccharides, some of the protein is needed to form the backbone of the foam on her beer. However, there does need to be generation of low-molecular-weight products, primarily amino acids, which the yeast will require for synthesis of its own proteins. Proteolysis is also necessary to get rid of proteins that contribute to haze formation in beer.

Barley contains a range of protein types, broadly classified by their solubility properties. In general, they can be divided into the water-insoluble proteins called hordeins, which are the major storage proteins, and into the water-soluble albumins, among which are the enzymes.

Proteolysis in the context of malting and mashing is primarily involved with the degradation of the hordeins. Two types of enzyme are involved. The proteinases attack these proteins in the middle of the molecule, releasing shorter linear polypeptide chains of amino acids. These shorter chains are the substrate for a second enzyme, called carboxypeptidase, which starts at one end of the chain, chopping off one amino acid at a time.

Carboxypeptidase is quite heat-resistant, but the proteinases aren't. Once again, then, brewers may start their mash at a lower temperature to deal with protein as well as β-glucan. This period of mashing is frequently referred to as a "proteolytic stand." However, there is increasing evidence that inhibitors that are extracted from the malt alongside the enzymes block much of the potential protein hydrolysis in a mash. Within the grain, the natural control mechanisms regulate the extent to which the inhibitors are able to interact with the enzymes. Once extracted, though, in what is literally a "mishmash," the inhibitors are freed from these restraints, and so proteolysis is limited.

ADJUNCTS

A Brewer may substitute a proportion of malt for various reasons, the substitute sources of extract being referred to as *adjunct*. Some adjuncts are used because they introduce necessary characteristics to a beer. For instance, the intense flavor of Guinness reflects the use of roasted barleys and malts in their grist. At the other extreme, some of the delicate character of Budweiser clearly originates in the rice that it contains, and the use of this material also allows for the product to have clean taste alongside a very pale color. Some Brewers will use adjuncts such as wheat flour because they believe they provide foam-enhancing substances to beer.

Sometimes, though, adjuncts will be employed for reasons of economy: if the unit cost of fermentable carbohydrate is lower in an adjunct than it is from malt, then it makes sense to replace a proportion of the malt, provided that it doesn't jeopardize any element of product quality, notably flavor. Some Brewers use corn products in the brewhouse. As we have seen, corn and rice starches have higher gelatinization temperatures than does barley starch, and they will need to be "cooked." Most commonly, hydrolyzed corn syrup or sucrose, both of which comprise ready-formed sugars that don't require an enzymatic stage in the brewhouse, will be used to supplement wort at the boiling stage in the kettle. In all cases, the brewer must perform his calculations carefully; if an adjunct is intended simply as a cheaper source of extract, it mustn't be forgotten that the additional processing costs for handling a more intransigent material may offset any potential savings. By using sugars and syrups in the kettle, the brewer can extend the volumes of wort produced without investing in extra mashing and lautering capacity.

THE BREWHOUSE

MILLING

Most frequently, malt is ground using roller mills, the malt being passed through one, two, or three pairs of rollers. The aim is to produce a particle distribution that is best suited to that particular brewhouse and the type of malt used. For example, if the husk of the malt is required as a filter bed for the separation of the wort, it will be necessary to have a setup that enables survival of this tissue, while at the same time milling the starchy endosperm to a fine enough consistency to allow easy access of water for its solvation. If the malt is relatively well modified, it will need less intense milling than would relatively undermodified malt to generate the same particle size distribution.

Generally speaking, the more rolls there are on a mill, the greater will be its flexibility. The brewer will inspect the milled grist, using a sieve to screen it into its various-sized components, and the roll settings will be adjusted if its particle distribution is felt to be suboptimal.

Some Brewers employ wet milling, in which the malt is steeped in water before milling begins. It is believed that the hydration of the husk lessens the risk of its damage during milling. Increasingly common is the use of hammer-milling, but only with modern mash separation processes such as the mash filter, which don't require the husk as a filter bed.

MASHING

This is the enzymatic stage of the brewhouse operation. The milled malt is mixed intimately with the water, which enables enzymes to start acting. Essential requirements of the stage are that the particles must be efficiently hydrated and careful control exerted over times and temperatures. It is by manipulating these that the brewer is able to influence the efficiency with which the malt is extracted.

Modern mashing vessels (still sometimes called *mash tuns* or *mash mixers*; see figure 8.1) are fabricated from stainless steel. This is the norm for all brewery vessels, as it makes for robustness and for ready cleaning by so-called Clean-in-Place (CIP) systems. To achieve intimate mixing of the milled grist and the water, they are mixed using a "fore masher" on their way into the mash conversion vessel. Rousers provide further mixing within the mash vessel. It isn't simply a matter of thrashing the mixture about. On the one hand, excessive physical damage to particles will slow down the subsequent wort separation stage and lead to unacceptably turbid worts, but it will also cause far greater uptake of air into the mash. It is now often said that this can promote staling in the subsequent beer.

Modern mash mixers are jacketed, with steam being used to heat up the contents of the vessel. As we have seen, mashing may commence at a relatively low temperature (say 45°C, 113°F) to enable the more heat-sensitive enzymes such as β-glucanase to do their work. When this "rest" is complete, steam will be put through the jacket to bring up the temperature to that required for gelatinization of starch.

Typical practice may be to introduce a proportion of water into the mash tun, sufficient to cover the agitator, before running in the grist/water mix via the premasher. Grist entry is likely nowadays to happen near the base of the vessel, so that air uptake is minimized. Various additions may be made. For instance, certain salts may be added if there is a need, for example, to adjust the chloride-sulfate balance. Calcium may be added in order to lower the

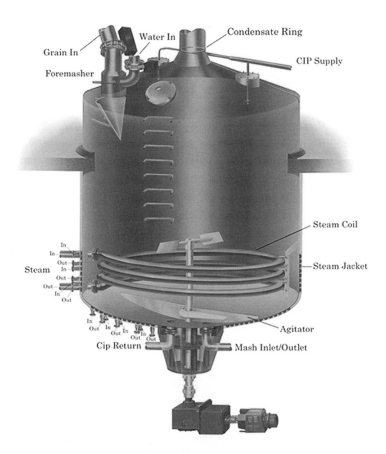

FIGURE 8.1 A mash mixer. The foremasher is where the grist and water are intimately mixed. The steam coil and jacket allow for the temperature to be raised in the vessel using steam. CIP signifies "cleaning in place," and this is the system used to wash and rinse the vessel between brews (courtesy of Anheuser-Busch)

pH of the mash (see Appendix): ideally, a mash should be of pH 5.2–5.6 for the appropriate balance to be struck between the various reactions that are occurring. Alternatively, acids may be used directly or introduced indirectly; for instance, in Germany lactic acid bacteria (so called because of their main excretion product) are used to acidify the mashes "naturally." Extra enzymes might be introduced in some countries, most often a heat-resistant β-glucanase to supplement the more sensitive enzyme from malt.

Cereal cookers used to gelatinize the starch in certain adjuncts are operated analogously to mash tuns, though of course the temperatures employed are higher.

WORT SEPARATION

Once the enzymes have completed their job in the mash, it is time to separate the resultant wort from the residual ("spent") grains. In many ways, this is the most skilled part of the brewing operation. The aim is to produce the kind of wort referred to as "bright": in other words, not containing lots of insoluble particles that may present great difficulties later on. The challenge is to achieve this without sacrificing wort, thereby limiting yields. Furthermore, this has to be achieved within a limited time period, for a brewer will want to put several brews through the brewhouse each day.

The majority of breweries in the world use a lauter tun (or tub) for this purpose (figure 8.2). In newer brewhouses, though, you are likely to find a mash filter.

The science of wort separation is fascinating, and is based on an equation developed by the nineteenth-century French engineer Henry Darcy:

$$\text{rate of liquid flow} = \frac{\text{pressure} \times \text{bed permeability} \times \text{filtration area}}{\text{bed depth} \times \text{wort viscosity}}$$

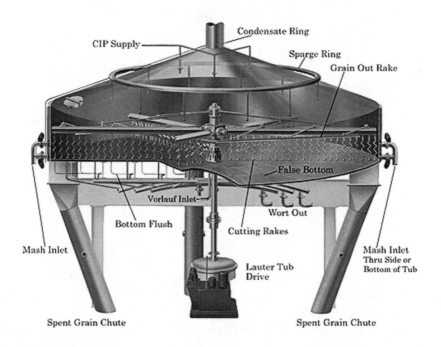

FIGURE 8.2 A lauter tun (courtesy of Anheuser-Busch)

Fundamentally, it means that the wort will be recovered more quickly if the vessel used to carry out separation has a large surface area and is shallow (i.e., the distance through the bed is short). Low viscosities (i.e., low β-glucan levels) will help, as will the application of pressure. The permeability depends on the particle characteristics of the bed. The best analogy would be to sand and clay. Sand comprises relatively large particles, whereas the particles of clay are far smaller. To pass through clay, water has to take a far more circuitous route than is the case for sand. Thus, big particles tend to present less of an impediment to liquid flow than do small ones. At the end of mashing and during sparging, relatively high temperatures (e.g., 76°C–78°C, 169°F–172°F) are maintained. In part, this is because of the inverse relationship between temperature and viscosity, but it is also known that smaller particles agglomerate to form larger ones at higher temperatures.

LAUTER TUN. Generally this is a straight-sided round vessel with a slotted or wedge-wire base and runoff pipes through which the wort is recovered. Additionally, within the vessel there are arms that can be rotated about a central axis. These arms carry vertical knives that are used as appropriate to slice through the grain bed and facilitate runoff of the wort. The brewer will first run hot water (at about 77°C, 171°F) into the vessel so that it rises to an inch or so above the false bottom. This ensures that no air is trapped under the plates, and it also serves to "cushion" the mash. The mash will then be transferred carefully from the mash tun to the bottom of the vessel, again to minimize oxygen uptake, and the knives will be used to ensure that the bed is even. Hot liquor is used to "rinse out" the mash tun and delivery pipes. The depth of the grain bed is unlikely to be more than 18 inches (see the Darcy equation above). After a "rest" of perhaps 30 minutes, the initial stage is to run off the wort from the base of the vessel and recycle it into the vessel, in order that it can be clarified. After 10 to 20 minutes of this so-called vorlauf process, the wort is diverted to the kettle and wort collection proper is started. This wort is at its most concentrated.

The remainder of the process is an exercise in running off wort that is as concentrated as possible within the time frame available. More hot (77°C, 171°F) liquor (the "sparge") is sprayed onto the grains so that the sugars and other dissolved materials are not left trapped in the spent grains. The knives are used as sparingly and carefully as possible so as not to damage grains and thereby make small particles that will clog the system or render the wort turbid or "dirty."

Another factor that the brewster must consider is the strength of the wort that is needed in the fermenter. If she is intending to brew a very strong beer, then clearly the wort must be rich in sugars. This limits the amount of sparge liquor that can be used in the lauter tun. Some brewers will collect separately in one kettle the initial stronger worts running off from the lauter tun, using this for stronger brews, before collecting subsequent weaker worts in a second kettle.

When the kettle is full, there may still be some wort left with the grains. Time permitting, this will be run off for use as "mashing-in" liquor for subsequent brews, a process referred to as "weak wort recycling." The brewer needs to be careful, though; when the worts are very weak, there is an increased tendency to extract tannins out of the grains, and these can cause clarity and astringency problems in beer.

At the completion of lautering, grain-out doors in the base of the vessel are opened and the cutting machinery is used to drive the grains out. The spent grains are trucked off site as fast as possible (they readily "spoil"), most commonly for direct use as cattle feed if there is a dairy or beef industry nearby.

MASH FILTERS. These operate by using plates of polypropylene to filter the liquid wort from the residual grains. Accordingly, the husk serves no purpose as a filter medium, and particle sizes are irrelevant. The high pressures that can be used overcome the reduced permeability due to smaller particle sizes (the sand versus clay analogy I used earlier). Furthermore, the grain bed depth is particularly shallow, being nothing more than the distance between the adjacent plates, which cumulatively amounts to a huge surface area. The chambers of the press are first filled with liquor, which is then replaced by mash with filling times of less than 30 minutes. During this time, the first worts are recovered through the plates. Once full, the outlet valves are closed. The filter is then given a gentle compression to collect more wort. This is followed by sparging to get a uniform distribution of liquor across the filter bed, then a further compression to force out the remaining wort. Using mash filters, wort separation can be completed in 50 minutes rather than the periods of up to 2 hours needed for lautering.

WORT BOILING

The boiling stage serves various functions. First, the intense heat inactivates any of the more robust enzymes that may have survived mashing and wort separation, and it sterilizes the wort, eliminating any organisms that might jeopardize the subsequent good work of the yeast. Second, proteins tend to coagulate when you heat them strongly, as anyone who has boiled an egg will appreciate, and so in wort boiling, proteins are removed that might otherwise precipitate out in the beer as haze. They cross-link with tannins (polyphenols) from malt and hops and produce what is known as "hot break." Third, the resins from hops are isomerized into the bittering principles, and other flavor changes take place, including the driving off of undesirable characters originating from hops and malt. The color of wort increases during boiling through melanoidin reactions (see chapter 4). Finally, as water is of course driven off as steam during boiling, the wort becomes more concentrated.

Most Brewers will tend to use a boil of between one and two hours, evaporating some 4% of the wort per hour. Clearly this is a most energy-intensive stage of the brewing process, and every effort is made to conserve heat input and loss. Kettles come in a myriad of shapes and sizes, but in modern breweries will be stainless steel, straight-sided and curved-bottomed, and very likely to be heated using an "internal or external heat exchanger" called a calandria (figure 8.3). Efficient boiling demands turbulent conditions in the vessel and thorough mixing; the calandria, which employs convective mixing of the system, enables this.

The significance of the boiling stage for beer flavor should not be underestimated. Apart from the driving off of unpleasant grainy characters that originate

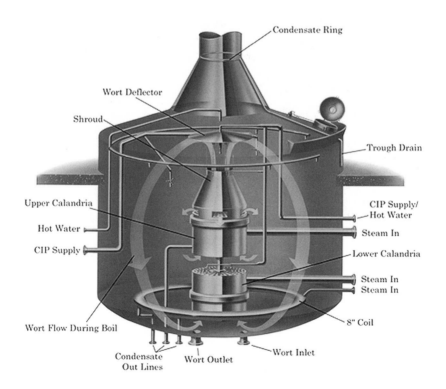

FIGURE 8.3 A brew kettle (courtesy of Anheuser-Busch)

in the grist, certain other substances are actually produced during boiling, and, at least in part, they can be desirable.

REMOVING TRUB

Various devices have been used to separate the trub and other residual solids from boiled wort. For those brewers using whole leaf hops, this stage is completed using a "hop back," analogous to a lauter tun, in which the residual plant material forms a filter bed. Such approaches are not applicable when hop pellets and hop extracts are used. Centrifuges have been used to remove wort solids, but much more common is the "whirlpool" or "hot wort receiver" (figure 8.4). These are cylindrical tanks, approximately 5 meters in diameter, into which wort is pumped tangentially through an opening between 0.5 and

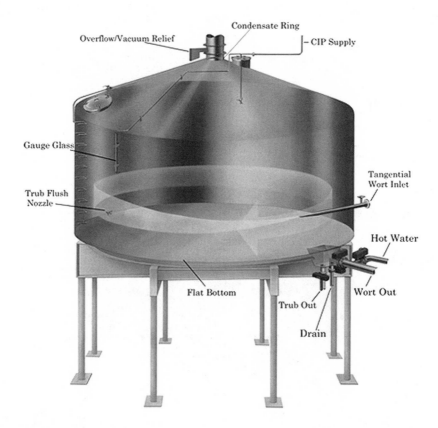

FIGURE 8.4 A whirlpool (courtesy of Anheuser-Busch)

1.0 meters above the base. The wort is set into a rotational flux, which forces the trub into a conical pile at the center of the vessel. After a period of up to one hour the wort is drawn off through pipes at the base of the vessel, so as not to disturb the collected trub.

The precipitation of insoluble materials in the brewhouse is sometimes promoted by the addition of a material called Irish moss, extracted from seaweed.

WORT COOLING

The whirlpool may be insulated, but if not, the wort temperature may fall to 85°C (185°F) or less. Even so, that is far too hot for the survival of yeast. For this reason, the final stage before fermentation must be cooling. Customarily, this is achieved using plate heat exchangers, which look like car radiators (figure 8.5). The wort flows turbulently on one side of the plates,

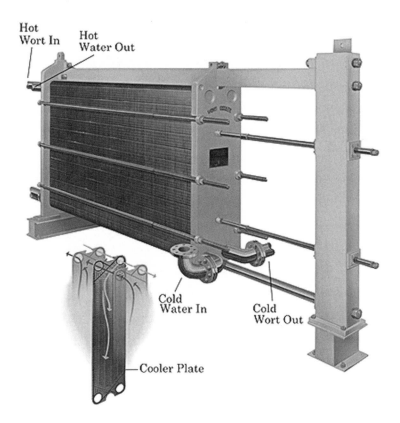

FIGURE 8.5 A Paraflow (courtesy of Anheuser-Busch)

with a cooling medium (chilled water, brine, ammonia, or glycol solution) on the other. Heat is convected across the plates from the hot wort to the coolant. When wort is chilled, more solids may precipitate out, the so-called cold break. Opinion differs between Brewers on the relative merits of this material (see chapter 7). Sometimes it is removed by flocculation, flotation, centrifugation, or filtration.

The final stage in wort production en route to the fermenters is the introduction of oxygen, which yeasts require for healthy growth, as we shall discover in the next chapter.

The sequence of events in the brewhouse, then, is complex, and they are all geared toward generating, in the highest possible yield, a nutritious wort that the yeast will grow on to make the beer the Brewer wants. The composition of that beer, and therefore whether it is good or bad, is inherently dependent on the behavior of the yeast, which in turn reflects the quality of the wort, as we shall see in the next chapter.

GODESGOOD

YEAST AND FERMENTATION

The common denominator to the production of all alcoholic beverages is fermentation. For beer this involves the conversion of sugars, derived primarily from malt, into ethanol (ethyl alcohol or, for most people, just "alcohol") by the yeast *Saccharomyces cerevisiae* or *Saccharomyces pastorianus,* the "mysterious" properties of which in medieval times caused it to be known as *"godesgood"* (God's good). In those days, nobody knew that they were dealing with a living organism; all they knew was that the barm collecting on top of a brew could be "back-slopped" into the next batch to allow it to get going rapidly.

The nature of any alcoholic drink is determined by the yeast strain used to produce it, but also on the substrate (feedstock) that the yeast is converting. Beers have the character they do because of the subtle interaction between carefully selected "brewing strains" of yeast and the malt and hops that come together as wort.

BREWING YEAST

Saccharomyces is a busy beast. Apart from being the workhorse of the brewery, it is responsible for the production of cider, wine, spirits, and some other alcoholic beverages. And as every cook knows, it is essential for the production of life's other staple food, bread.

The reader needs to be aware that it is not the same strains of yeast that do all these tasks. Just as it takes humans with all manner of skills to make up society, so is it a collection of strains of *Saccharomyces* that tackle the range of tasks referred to above. Yes, brewing strains can be used to ferment grape must and make passable wine, and wine yeasts can be used to ferment wort with some interesting products. The fact remains, though, that the character of a beer is in large part established by the yeast that is used to make it. That is why Brewers guard their brewing strains carefully—just as any skilled workman looks after the tools of his trade.

THE STRUCTURE OF YEAST

Yeast is a single-celled organism, about 10 millionths of a meter in diameter (figure 9.1). Bacteria also consist of a single cell, but yeast is substantially more complex, and, like all so-called eukaryotic organisms, the cell is divided up into organelles, each with its own job of work. Yeast is actually the "model" organism used in the basic laboratory studies that ultimately project to findings on more complex eukaryotes, especially the human.

The heart of a cell is its nucleus, within which is stored much of the genetic information held in deoxyribonucleic acid (DNA). In turn, the DNA is coiled up into sixteen different chromosomes. The strains of this organism that have been used for much of the laboratory research over the years contain just one or two copies of each chromosome: they are said to be haploid or diploid. (Most of your cells, dear reader, and mine are diploid—one set of chromosomes from each of our parents.) Brewing yeasts are aneuploid, containing approximately three copies of each chromosome. I say "approximately" because the exact number of copies of individual chromosomes present may differ. The fact that brewing yeasts contain more than a single copy of each gene makes them quite stable; they can tolerate loss of one of the copies of a gene because they can fall back on the other copies. This is good news for Brewers, as their yeasts are consistent for many generations.

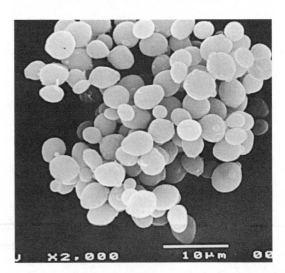

FIGURE 9.1 Brewing yeast
(courtesy of Katherine Smart)

The yeast cell is surrounded by a wall, within which is a membrane. The wall offers strength to the cell, protecting the rather more delicate membrane beneath it. It also plays a major role in cell-cell interactions: it is through links between walls and calcium that cells flocculate and migrate either to the liquid surface or to the bottom of a fermenting vessel. This has major implications for brewing practice, for instance the procedure that the brewer will use to separate the yeast from the "green" beer at the end of fermentation.

The function of the membrane is to regulate what does and what does not get into and out of the cell. Although a yeast has its intracellular food reserves, it depends on materials in its growth medium (in the case of beer, the growth medium is wort) for its survival and growth. The composition of the membrane influences what (and how readily) molecules such as sugars and amino acids move into the cell. The membrane has a similar influence on what leaves the cell.

Like other single-celled organisms, brewing yeast reproduces by cell division. The daughter cell grows from the mother cell as a bud before separating

---

BOX 9.1 GETTING HOLD OF YEAST

I am fond of telling my students that the best way to obtain a new yeast strain is to steal it. I tell them of an erstwhile boss of mine who supposedly visited a Czech brewery and collected a nice sample by surreptitiously swiping his handkerchief across the rim of an open square fermenter and delivering said piece of linen back to home base for our bug hunters to isolate the prized beast (which had an above-average chance of having actually come from up the great man's nose).

History tells us that the origin of modern lager yeasts is in a Bavarian monastery-brewery of 300 years ago, with the organism subsequently spirited away by brewers from a famed Eastern European brewery country and then in turn stolen from them by a representative of a company in an equally famous northern country. I would, of course, never venture to comment on the rights and wrongs in any of this, save to say that if latter-day production strains are very largely subtle variants of this "pilfered" predecessor, then it's small wonder that I downplay the significance of strain.

Of course, you don't have to obtain yeast by cloak and dagger. For example, you can screen a wide assortment by seeking out strains held by time-honored culture collections such as the National Collection of Yeast Cultures (NCYC, http://www.ifrn.bbsrc.ac.uk/ncyc/) in the United Kingdom or the American Type Culture Collection (ATCC, http://www.atcc.org/) in the United States. Holland has Centraalbureau voor Schimmelcultures (www.cbs.knaw.nl), while in Belgium there is the Belgian Co-ordinated Collections of Micro-organisms (www.belspo.be/bccm/). And closer to my home is the collection of the late, great taxonomist Herman Phaff, expertly looked after these days by Kyria Boundy-Mills (http://www.phaffcollection.org/).

off as a distinct cell, leaving a "bud scar" behind on the mother cell. Figure 9.1 shows yeast viewed under an electron microscope; budding cells and bud scars are readily seen. An indication of the age of a yeast cell is obtained by counting the number of bud scars, which can be as many as forty or fifty. Yeast can also enjoy a healthy sex life, though perhaps sadly for it this is hardly a daily occurrence.

CLASSIFICATION OF BREWING YEASTS

Brewing yeasts are divided into two categories: *Saccharomyces cerevisiae* and *Saccharomyces pastorianus* (the latter was previously called *Saccharomyces carlsbergensis* and then *Saccharomyces uvarum*). These are respectively ale yeast and lager yeast. The latter do their job at relatively low temperatures, typically between 6°C and 15°C (43°F–59°F), and after flocculating they drop to the bottom of the fermenter; they are traditionally used in the production of lager-style beers. *Saccharomyces cerevisiae* is used to make ales at temperatures in the range 18°C to 22°C (64°F–72°F); this type of yeast collects at the surface of the fermenting vessel. It is, indeed, possible to differentiate between yeasts in either category, notably by the fact that yeasts classified as *S. pastorianus* can grow on the sugar melibiose, whereas *S. cerevisiae* can't. In fact, it seems that *S. pastorianus* evolved from *S. cerevisiae* by melding with a different yeast, *Saccharomyces bayanus,* the organism that is used by winemakers for the production of certain sherries and in the making of champagne.

Because brewing strains differ so much in their properties and behavior, it is important that a brewer know which strain he is dealing with. For instance, a company may brew the same brand in several different breweries and distribute the relevant yeast from a central repository. The sender and the recipient should both run checks to make sure that the yeast is the right one. Within a given brewery, too, several yeast strains may be used to make a range of products. It is critical to be able to distinguish them. Good housekeeping only goes so far; from time to time, a check needs to be run to confirm that the correct yeast is being used. This problem is particularly acute where a brewery performs franchise brewing. I know of one major brewery, for instance, that brews at least four major international lager brands for four different companies. Not only is that brewery trusted with custodianship of the respective yeasts, they are also under intense pressure to make sure that there are no mix-ups or cross-contaminations.

There are those who downplay the significance of yeast strain, and, in fact, there is clear evidence that certain brands can be successfully made with yeasts associated with a totally distinct brand. (I know of one famous brewer who used to use baker's yeast.) Indeed, there are opportunities for rationalization of yeast strains—but this demands rigorous trials to ensure that the desired beer is produced (and will continue to consistently display the required quality characteristics) and that there are no "funnies" in production. Such rationalization is far easier to achieve for the brands within a company. By and large, Brewers who franchise out a brand demand that their specified process be adhered to, using their specified raw materials—and that includes yeast. Fundamentally, the greater the contribution of grist and hops to the flavor delivery of a beer, the less is likely to be the significance of the yeast strain to the character of that beer.

Identification of brewing yeast strains was once performed using a battery of physiological tests. Nowadays "typing" is achieved using DNA fingerprinting, in a technique exactly analogous to that employed in a criminal investigation. One might almost say that the "rogue" under pursuit is the yeast strain different from that which should be used to make the beer brand required.

THE USE OF WORT BY YEAST

Like any other living organism, yeast needs certain essentials to enable it to grow and survive. It needs vitamins (though later in fermentation it will synthesize these important substances itself), it needs a source of nitrogen (amino acids from the breakdown of barley protein during malting and mashing) that it will use to make protein, and it needs a few trace elements. Above all else, yeast requires sugars, which it will burn up to release energy and to make smaller molecules that it will use alongside the nitrogen source to fabricate its cellular components.

Yeast can use sugars in one of two ways. If it encounters high levels of sugar, such as are found in wort, it will use them by a fermentation (anaerobic) process. They are converted into ethanol and carbon dioxide, with the release of energy:

$$C_6H_{12}O_6 \rightarrow 2C_2H_5OH + 2CO_2 + energy$$
$$\text{sugar} \qquad \text{alcohol}$$

However, if the sugar content is low and if oxygen is available, then the sugar is used by respiration:

$$C_6H_{12}O_6 + 6O_2 \rightarrow 6CO_2 + 6H_2O + energy$$

In fermentation, about fourteen times less energy is captured for each glucose molecule broken down than is the case in respiration—but it's still enough for the needs of the yeast, because of the high availability of its sugar feedstock. If the cell has lots of sugar on which to gorge itself, it has no need to do extra work making all the extra enzymes that it will need to perform respiration.

This biochemistry is the basis of differentiation between yeast operating in a brewery and yeast being produced commercially for use in baking. In the latter case, it is economically desirable to produce large amounts of yeast from as little sugar as possible. Therefore yeast is grown in a so-called fed-batch process in which the sugar source (usually molasses) is dosed in a bit at a time, so that at any point its concentration is low and the yeast is switched on to using it by respiration. Plenty of oxygen is introduced, and the yield of yeast is high. A Brewer, on the other hand, is interested in alcohol production. Sugar concentrations therefore are high in wort, oxygen levels are low, and the yeast metabolizes the sugars by fermentation. Indeed, the Brewer wants very little yeast production, because the more sugar ends up in new yeast cells, the less has been converted into alcohol.

Even for the brewing of beer, yeast needs a little oxygen. Earlier we saw that the yeast membrane is important for healthy yeast. This membrane contains various components, among which are lipids. Yeast uses oxygen in the synthesis of these materials. So the brewer carefully introduces just the right amount of oxygen into wort to enable the production of the appropriate amount of membrane material. Too little, and the yeast won't ferment the wort efficiently. Too much, and yeast growth will be excessive and alcohol yield will be lowered.

SETTING UP A BREWERY FERMENTATION

Getting the right level of oxygen into wort prior to pitching the yeast is but one of the conditions that has to be right.

First, the wort itself. It needs to have the correct strength in terms of level of sugar. Increasingly, fermentations are performed at so-called high gravity, in which case the concentration of wort (and, proportionately, of oxygen and yeast) is higher than needed to give the desired final alcohol content. This is to maximize fermenter capacity. At the end of the process the beer is diluted with deaerated liquor (water) to the required alcohol content.

Irrespective of whether fermentation is at high or at "sales" gravity (i.e., fermentation of wort at the strength that gives the required beer without dilution), the concentration of sugars is measured by specific gravity (which is the weight of a volume of the wort relative to the weight of the same volume of water). So a wort may have a specific gravity of 1.040, which means that 1 milliliter of it weighs 1.04 g (1 mL of water weighs 1 g). Most brewers actually quote wort strength as "percent Plato," 1% Plato equating to 1 g sucrose per 100 g water. So, if wort has a specific gravity of 10% Plato, it has the same specific gravity as a 10% solution of sucrose.

The wort also needs to have the required level of solid material suspended in it. This is the so-called cold break produced in the brewhouse (see chapter 8), which is rich in lipids. Brewers differ hugely in their opinions on whether the presence of this material is a good or a bad thing. Some, for instance many German brewers, are adamant that cold break causes only problems, and that it is a serious cause of poor foams and excess staleness in beer. The converse view is that some solids in wort are good news, because they promote a vigorous fermentation by acting as a site where carbon dioxide bubbles form, thereby releasing the yeast from potential "suffocation" by excessive levels of this gas. The bubbles also have a buoyancy role, keeping yeast in suspension and therefore in contact with wort for sustained and efficient fermentation.

The next essential is to use the correct level of yeast that is in the proper state of health and purity. The process of adding yeast to wort is called "pitching." As a rule of thumb, 1 million yeast cells will be added for every milliliter of wort and every degree Plato.

To measure the amount of yeast, most brewers will count the number of cells seen under a microscope in a drop of the yeast suspension placed on a special slide that is divided into grids. This device is called a hemocytometer, because it was originally developed for counting red blood cells in clinical labs. By knowing how much suspension was put onto the slide, the brewer can calculate the cell concentration. Some Brewers are rather more sophisticated than this and automatically dose yeast on the basis of measurements made with probes put directly into the pipeline that leads from the yeast storage tank to the fermenter. These probes work on various principles, among which are measurement of the capacitance of the yeast cells and light scatter. Suspensions of particles, such as yeast, scatter light in proportion to the number suspended per unit volume. The other method takes advantage of the fact that yeast cells can store electrical charge (i.e., they are capacitors) in proportion to how many cells are present. Dead cells and trub do not register.

The number of yeast cells added is important. So too is the health of the yeast. Dead cells won't ferment wort into beer. Just as significantly, the products of their decay can cause problems for the brewer. The most common means of measuring the viability of yeast involves a dye called methylene blue. Living yeast is capable of decolorizing this dye, but dead cells aren't, and as a result, they stain blue.

Even if a cell is living, that doesn't necessarily mean that it is in a fit state for carrying out an efficient fermentation. When yeast is in a healthy and vigorous condition, ready to do its job, it is said to have vitality. The analogy would be a comparison of the average couch potato watching a ball game on television with a major league baseball star, perhaps a Tim Lincecum. Both are living, but it is the latter who possesses vitality and who is, well, better suited to *pitching*. Measurement of vitality is not a straightforward issue, and there is no agreement on the best way of assessing it. Most brewers recognize that the most appropriate course of action is to look after their yeast and ensure that it doesn't encounter stresses such as heat shock or those that arise from leaving it in contact with beer long after fermentation is complete. By protecting the yeast, they stand to keep it in good condition.

The only other ingredients likely to be included into a fermentation are a "yeast food," most frequently a zinc salt, and antifoam. Zinc is a key component of one of the enzymes that yeast requires to carry out alcoholic fermentation. Other yeast foods are more complex mixtures of amino acids and vitamins, but many folk would have it that this solid addition merely acts as a nucleation site in just the same way as cold break.

Antifoam is required if a fermentation is characterized by high levels of head formation. This occurs particularly with certain types of yeast and for fermentations carried out at the higher end of the temperature range. Such "overfoaming" has two consequences. First, the capacity of the vessel is reduced: the brewer is obliged to put less wort into the fermenter, otherwise it will overflow during the process. Secondly, any foaming during the process reduces the amount of material that will survive to support the head on the finished beer in the glass. To minimize this foaming, many brewers add antifoam agents, most frequently those based on silicone. It is essential that they are removed efficiently by adsorption onto the yeast and in the filtration operation, otherwise they will damage the head in the beer itself. In a country such as the United States, antifoams tend to be avoided. Indeed, this preference for not using additions extends to many other materials that are employed in other parts of the world as a matter of routine throughout malting and brewing.

THE FERMENTATION CELLAR

Many types of fermenter exist in breweries across the world. Common types are square or rectangular and cylindroconical (figure 9.2). The original commercial fermenters were "open squares," and these are still used extensively for the production of ales in the United Kingdom. These days they are fabricated from stainless steel, but through the years they have been constructed from oak, slate, copper, and reinforced concrete. They come in a vast range of sizes, and cylindroconical vessels capable of holding over 13,000 hectoliters have been used. More commonly, "squares" are between 150 and 400 hectoliters. Squares are highly suited to fermentations with top-fermenting yeasts, with the yeast periodically skimmed from the surface of the vessel. Clearly, there is some risk of contamination, and you will soon discover if you lean over such a tank and inhale that there are vast quantities of carbon dioxide evolved that will, literally, take your breath away. (Incidentally, in case the reader was worrying that the Brewer is carelessly pumping greenhouse gases into the atmosphere, it should be stressed that the amount of carbon dioxide produced by fermentation in the world's breweries is very minor when compared with the amount of this gas that you and I and the rest of the world's animal population breathe out every second of every day. Not only that, but remember that it takes a lot of carbon dioxide to support the growth of barley and hops by photosynthesis—rather more, in fact, than is produced during fermentation of beer.)

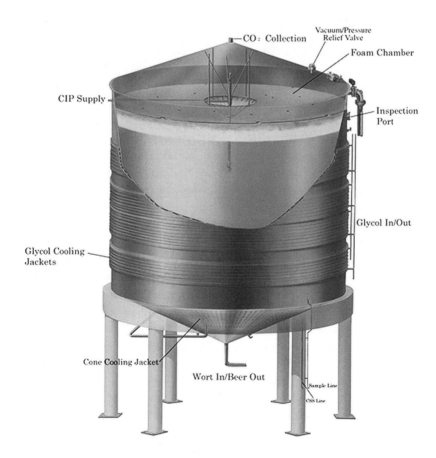

FIGURE 9.2 A cylindroconical fermenter (courtesy of Anheuser-Busch)

Many Brewers seek to collect the carbon dioxide from fermenters to put, for instance, into cylinders and use as a motor gas in pub dispense systems. CO_2 collection is possible from closed fermenters. Some of these vessels are little more than open squares with a lid on, but for the most part fermenters these days are cylindroconical tanks, which are seldom of a capacity less than 600 hectoliters but can be as large as 7,000 hectoliters. (Some brewers use tanks that are similarly sized but horizontal—essentially, cylindrical tanks laid on their side.) Once, the trend was for installation of bigger and bigger vessels, and indeed, such vessels do make sense in breweries that have limited ground space and are producing large volumes of one or a very few brands. There are, however, potential problems, insofar as yeast does behave differently depending on the hydrostatic pressure it encounters, and it may change its output of flavor materials, leading to a perceptibly different character in the beer. In particular, though, for breweries producing a diversity of brands it makes more sense to use smaller fermenters.

Cylindroconical vessels were originally developed by the Swiss Leopold Nathan at the turn of the twentieth century and have the advantages of better mixing due to convection currents set up by rising gas bubbles, ease of temperature control through thermostatted jackets, and easy and hygienic recovery of yeast from the base (cone). These vessels are also easily cleaned using a water spray, followed by either dilute (1%) caustic or phosphoric acid and another water rinse, usually prior to a sterilant rinse with either hypochlorite, chlorine dioxide, or perchloric acid. These various treatments are sprayed into the empty tanks from a spray ball (nozzle). (Incidentally, such Clean-in-Place, CIP, is also employed at other stages through the brewery between brews to ensure cleanliness in all types of vessel and pipeline.)

It is possible to deliberately apply a pressure to these vessels during fermentation; the formation of esters, for instance, is suppressed at higher pressures.

Whichever type of fermenter is employed, the principles of what happens during the fermentation are similar. Yeast takes up sugar (and the other materials) from wort and converts it into alcohol and CO_2. Most commonly, the progress of fermentation is monitored by measuring the decline in the specific gravity of the wort. This decrease occurs because the specific gravity of a solution of ethanol is vastly lower than that of a mixture of sugars. Alongside the fall in specific gravity is a drop in pH, as yeast secretes hydrogen ions and certain organic acids (such as citric and acetic acids, the acids found in lemons and vinegar, respectively—happily there is rather less of either in beer) and also digests materials from the wort that act as buffers (materials that are capable of holding pH constant).

During the fermentation a range of molecules leaks out from the yeast cell, among which are substances that have distinctive flavors. They include esters and higher alcohols (which collectively are sometimes referred to as fusel oils), certain sulfur-containing molecules, and a particularly noxious material called diacetyl that has a distinct butterscotch character (see chapter 4).

Typically lagers will be fermented at temperatures between 6°C and 14°C (43°F–57°F), with the chosen temperature being controlled very carefully by the brewer. Generally speaking, the more traditional the Brewer of lager beers, the lower will this temperature be. Clearly, rates of fermentation are slower at lower temperatures. This leads to a different balance of flavor substances released by the yeast. The traditionalists would contend that the best flavor balance is achieved if the process is painstaking—at lower temperatures. Others insist that perfectly good beer is produced by fermentation at the higher end of this temperature range. Such differences of opinion mean that fermentation of lager can take as little as 3–4 days, but as much as 2 weeks.

Ales have always been fermented at higher temperatures (15°C–20°C, 59°F–68°F) than lagers, with the result that they tend to contain more flavor volatiles, such as esters, than do lagers. These fermentations also tend to be faster.

The vast majority of Brewers agree that diacetyl is an undesirable substance to have in the beer. This substance leaks out of yeast during fermentation, but is subsequently taken up again by the yeast at the end of fermentation. The process must be continued until the diacetyl has been lowered to below 0.01–0.1 mg per liter (the target differs between Brewers—the "gentler" the flavor of the beer, the lower the target is), and this depends on there being enough healthy yeast present at the end of fermentation. Some Brewers allow the temperature to rise at the end of the primary fermentation to allow this mopping-up operation to occur more rapidly. Others practice krausening, in which a charge of freshly fermenting wort with a very high count of vigorously growing yeast is added late in the fermentation/maturation to provide an abundance of cells capable of eliminating the diacetyl.

Once fermentation and diacetyl removal are complete, yeast is separated from the beer. If the two are allowed to remain in contact for too long, materials can leak from the cells that can damage the beer. As we have seen, much of the yeast can be readily separated from the beer, either by skimming, in the case of a top-fermenting strain working in an open square, or through the collection of a bottom yeast in the cone of a cylindroconical vessel. If any further help is required, it comes in the form of a centrifuge.

There are three possible fates for the yeast. It can go to a chilled storage tank for holding a few days prior to pitching into another fermentation. Some brewers collect the yeast in a press. Alternatively, it may be used immediately for pitching into another vessel; this practice is often called cone-to-cone pitching, as it involves the pumping of yeast from the cone of one vessel in which fermentation is complete into the cone of a vessel containing fresh wort. The third option is for the yeast to be disposed of. A proportion may go off to a distiller for whisky fermentation. Some will be treated with propionic acid prior to ending up as pig slurry. The majority in the United Kingdom and Australia, though, goes off for autolysis and making into yeast extracts marketed as Marmite or Vegemite that will end up spread on somebody's toast! It's a taste to be acquired at a very early age—the manufacturers even advertise Marmite with the slogan "you either love it or hate it"!

Storage of yeast prior to repitching is itself a process demanding great care. The yeast is kept well mixed (roused), and a little oxygen may be introduced to keep the cells ticking over. The tank is also likely to be thermostatted to 0°C–4°C (32°F–39°F). The yeast may also have picked up some contaminants in the fermenter, which must be gotten rid of. This can be achieved by washing the yeast for an hour or two in a very cold solution of phosphoric acid. Healthy yeast survives this treatment quite happily, but bacteria don't.

YEAST PROPAGATION

Some Brewers have kept the same yeast going for years and years. Some of the yeast produced in fermentation is used to pitch the next batch of wort, and so on. However, it is a fact that the yeast genome (despite the aneuploidy mentioned above) is not stable, and it is desirable to repropagate each yeast strain after every five to ten batches of wort have been fermented.

Propagation is from stock yeast that may be held in various ways but is increasingly likely to be either a deep-frozen culture or one that has been freeze-dried. When it is time to propagate, this culture will be used to inoculate a small amount of wort (perhaps 10 ml), with growth of the yeast being in progressively increasing amounts of medium (100 ml, 1 liter, 5 liters, until the final propagator, which may have vessels ranging in capacity from 10 to 300 hectoliters). Rigorous conditions of sterility are essential, as is a plentiful supply of sterile wort and oxygen. The aim of propagation is to produce large quantities of yeast that is in good condition for subsequent brewery fermentations. Because respiration yields far more energy than does fermentation and ethanol is not a desired product of propagation whereas yeast biomass is, highly aerobic conditions should be maintained in a propagator. The most efficient way to grow yeast is in fed-batch mode, which the purveyors of baker's yeast have long since appreciated (see above).

WHAT YEAST EXCRETES

Beer is, of course, a delicious and wholesome product. The fact remains, though, that it is merely the spent growth medium of a fermentation process.

Beer is the way it is because of the things that yeast takes away, the substances that is excretes, and the stuff that it leaves well alone. Yeast "eats" sugars, taking away excessive sweetness while simultaneously producing its most significant excretion products, ethanol and carbon dioxide. It doesn't metabolize the proteins or the bitter compounds, although both can adsorb onto the yeast wall. And, importantly for the flavor of beer, yeast releases flavor compounds.

We have already heard about diacetyl, which is extremely undesirable, with the remotely possible exception of the very occasional ale where a low level might have some benefit. Two categories of substance that are desirable when present in the appropriate quantity for a given beer are the higher alcohols and more particularly their equivalent esters. The levels obtained in beer depend on fermentation conditions: levels increase at higher fermentation temperatures if less yeast is pitched into the fermenter and if insufficient oxygen is used. Increasing the top pressure in fermentation can suppress the tendency of esters to be produced. Of particular significance, however, is the yeast strain, some strains producing more esters than others.

Yeast secretes a range of sulfur-containing compounds into beer, including hydrogen sulfide, dimethyl sulfide, and sulfur dioxide. Sulfur dioxide is produced by yeast from the sulfate present in wort and also from some of the sulfur-containing amino acids. While not itself as flavor-active as other sulfur compounds, sulfur dioxide can suppress the deleterious flavors due to other compounds that can arise in beer. Notably, sulfur dioxide acts as an antioxidant and helps prevent stale flavors from developing in the product.

By and large, the other sulfur compounds present in beer are strongly flavored at extremely low levels. Despite their individual pungency, if they are present at relatively low levels and in the correct balance, they contribute beneficially to the flavor of many beers, especially lagers. As for other products of yeast metabolism, there are substantial differences between yeast strains in their ability to form the various sulfur compounds. A major factor, therefore, in controlling the flavor of beer is to ensure that you use the correct yeast strain, and only when it is in good condition.

The immediate precursor of ethanol, acetaldehyde, is another potent flavor compound that, if present, gives a green-apple flavor to beer. Ideally it shouldn't be present, but if too much oxygen is present during fermentation then it can occur. It can also be symptomatic of the presence of spoilage organisms, in this case *Zymomonas*. Indeed, abnormal levels of other flavor constituents of beer, including some of the sulfur compounds, may also be due to infection.

Organic acids (including succinate, lactate, and acetate) are normal products of the metabolism of brewing yeast. Their secretion contributes to the characteristic fall in pH that occurs during fermentation, from over 5.0 to as low as 3.8. Finally, yeast can produce medium-chain-length fatty acids, such as octanoic and decanoic acids, which can provide flavors to beer such as "goaty" and "wet dog."

MODERN FERMENTATIONS

Traditionally, fermentation was performed at "sales gravity"; in other words, the strength of the finished beer was in direct proportion to the concentration and the fermentability of the sugars in the wort. This is still the norm for many Brewers, particularly those producing smaller volumes of beer. These Brewers are more likely, too, to adhere to other traditional elements of the fermentation process, such as low temperature and fermentation at atmospheric pressure. Other Brewers, meanwhile, have considered and, in many cases, implemented procedures that will greatly enhance the productivity of their plant.

HIGH-GRAVITY FERMENTATIONS

Many Brewers perform their fermentations at concentrations of wort that give alcohol yields in excess of target. Following fermentation and conditioning, the beer is diluted to the specified alcohol content by the addition of water (which must be deaerated to prevent oxidative damage to the beer and preferably carbonated to the level of the beer it is diluting). Thus, for a beer that might traditionally have been fermented from a wort of 10°P to give 4.5% alcohol, in high-gravity fermentations the yeast might be pitched into 16°P wort, and the ensuing beer of 7.2% alcohol diluted 10 parts beer to 6 parts deaerated liquor to produce the desired final beer strength.

Commercially, 20°P appears to have been as high as anyone has successfully fermented high-gravity brews. Provided sufficient fermentation and downstream facilities are available, it will be seen that high-gravity brewing presents tremendous opportunities for enhancing brewery capacity and maximizing the amount of beer produced per unit of expenditure on items such as energy. To be successful, of course, it is essential that there is the wherewithal to produce such concentrated worts; furthermore, sufficient control must be exerted to ensure that the finished beers are indistinguishable from those produced at sales gravity. High-gravity worts going to fermenter can be produced by mashing at lower water-grist ratios, restricting such worts to the concentrated flows emerging early in the wort separation stage (see chapter 7), or, most typically, by boosting the levels of fermentable sugar by adding syrups to the kettle boil. Several problems must be overcome. Hop utilization is inferior at higher wort strengths; brewhouse yields are, of course, poorer; and yeast behaves differently when confronted with extra sugar, finishing the fermentation in a less healthy condition and producing disproportionately high levels of certain flavorsome substances, notably esters, as well as releasing enzymes that damage foam. These problems are not insurmountable, and the combined use of higher yeast pitching rates and proportionately more oxygen for the yeast to use for membrane synthesis means that large quantities of the world's beer are now produced most successfully in this way.

ACCELERATED FERMENTATIONS

Another way to enhance capacity would be to increase the turnover of fermenters, that is, to speed up fermentations. This can be achieved by increasing the quantity of yeast pitched into fermenter (with oxygen enhanced proportionately), maintaining yeast in contact with the wort rather than allowing it to flocculate, and by elevating the temperature. In each case there is invariably an effect on flavor, which will need to be addressed, perhaps by increasing the top pressure on the fermenter if this is feasible.

CONTINUOUS FERMENTATION

Many industrial fermentations are performed continuously. With a solitary exception, this is not the case for brewery fermentations, despite the obvious potential advantages for turnover and capacity. At times over the past thirty years, various breweries did install continuous fermentation processes, notably employing tower fermenters with upflow of the liquid stream through a heavily sedimentary yeast capable of forming a plug at the base of the vessel. By adjustment to the yeast content and the rate of wort flow, green beer could be produced in less than a day. With one exception, these fermenters have been stripped out, the main reasons given being inflexibility (most breweries produce a range of beers that demand diverse fermentation streams) and infection problems: it's bad enough having a contamination in a batch fermenter, but substantially more inconvenient if the fermentation is continuous. There is also the matter of beer flavor; it is an undeniable truth that virtually any change in fermentation conditions, whether temperature, yeast concentration, or in this case continuous processing, leads to flavor shifts.

These problems are certainly not insurmountable—as has been proved by Dominion Breweries in New Zealand, who for many years have used continuous fermentation to produce some prize-winning beers. Indeed, there is a resurgence of interest from others in this area, including the use of so-called immobilized yeast, where the yeast is attached to a solid support and the wort is flowed past. One Dutch Brewer employs this type of process in the production of a low-alcohol beer, while others (notably in Japan) are experimenting with such fermentation systems for making full-strength beers on a boutique brewery scale. Furthermore a Brewer in Finland employs immobilized yeast in an accelerated process for eliminating diacetyl at the end of fermentation.

Fermentation is now done, and the contemptible diacetyl destroyed, but the brewer's job is far from over. The "green beer" produced still needs to be refined in terms of its flavor and its appearance. Chapter 10 will tell us how that is achieved and how the beer is sent into the marketplace.

REFINING MATTERS

DOWNSTREAM PROCESSING

When a beer leaves the fermenter, it is not the finished article. It is highly unlikely to be sufficiently clear ("bright") and will certainly contain substances that will come out of solution in the ensuing package. Its flavor may still require some refining. All Brewers recognize the need to attend to the "raw" or "green" beer, but they differ in their opinions about quite how intense and involved this processing needs to be.

FLAVOR CHANGES DURING THE AGEING OF BEER

As we saw in chapter 8, a time-consuming step for moving beer onward from the fermenter is the time taken to mop up diacetyl and its precursor. Some people refer to this as "warm conditioning." Many Brewers would consider this to mark the end of the useful flavor changes that they can dictate in the brewery. The traditionalists would contend that the beer still needs to be stored. There is, however, very little published data to indicate what, if any, further changes take place in the flavor of beer when it is aged in the brewery.

Some major brewing companies insist on holding lager for a prolonged period at low temperatures (decreasing from 5°C to 0°C, or 41°F to 32°F). This process (lagering) is a leftover from pre-refrigeration days, when the removal of bottom-fermenting yeast demanded that the beer be held for a long time, with chilling perhaps facilitated by blocks of ice. Traditionally, beer from an already relatively cool fermentation (<10°C, 50°F) was run to a cellar at a stage when there was still about 1% fermentable sugar and sufficient yeast left in it. The yeast would consume traces of potentially destabilizing oxygen and, by fermenting the sugar, release carbon dioxide that would remain in solution to a greater extent at the lower temperatures and "naturally" carbonate the product. In this way the beer might be held at 0°C (32°F) for perhaps fifty days. Yeast would settle out by the end of this time, together with protein and other material that otherwise would "drop out" as an unsightly haze in the finished beer in the customer's glass. Adherents to the technology insist that subtle changes occur in the balance of flavor compounds in the beer, in particular the removal of undesirable notes such as acetaldehyde.

These days, the technology exists to cover all these requirements for prolonged storage, including the use of clarifying agents, filters, stabilizing agents, and carbonation systems, all allied to the use of refrigeration, as we will see in this chapter. This doesn't stop some major players in the brewing world insisting on the costly process of holding beer in tank for many days. They are convinced it is right, and as one of them so famously remarked: "If it ain't broke, don't fix it."

THE CLARIFICATION OF BEER

COLD CONDITIONING

Two types of particle need to be removed from beer at the end of fermentation: yeast and cold break. In addition, substances that are present in solution at this stage but that will tend to form particles when beer is in the trade must also be eliminated. We'll come back to that later.

The first mechanism by which particles separate from beer is simple gravitational pull. Most Brewers ensure that their beer is chilled to either 0°C (32°F) or, better, –1°C or –2°C (30°F–28°F) after it has enjoyed the degree of fermentation and maturation that they deem it requires. Particles will progressively sediment at this temperature in proportion to their size, and furthermore, materials will be brought out of solution, substances that might otherwise emerge as unsightly haze in the packaged beer.

To facilitate the sedimentation of particles, many Brewers add isinglass finings. These are solutions of collagen derived from the swim bladders of certain species of fish from the South China Sea, the dried bladders being referred to with such colorful titles as Long Saigon, Penang, and Brazil lump. Collagen has a net positive charge at the pH of beer, whereas yeast and other particulates have a net negative charge. Opposite charges attracting, the isinglass forms a complex with these particles, and the resultant large agglomerates sediment readily. Sometimes the isinglass finings are used alongside "auxiliary finings" based on silicate, the combination being more effective than isinglass alone.

BOX 10.1 ISINGLASS

One of the great beer genres, the English cask ale, emerged on the backbone of a "natural" clarification process rooted in a protein preparation called isinglass. It is obtained from the dried swim bladders (some call them "maws") of certain warm-water fish, among them the catfish, jewfish, threadfish, and croaker.

Lest you were concerned that the greedy isinglass folks whip out the swim bladder and toss the rest of the beast overboard, let me assure you that these fish are primarily caught for food use. The functional property of the maw is sublime added value. In any event, the bladder is even more likely to end up in a Hong Kong soup than as finings for the beer and wine industries.

The bladders are removed, washed, and dried. At the smallest scale, in a fishing village, the maws are sun-dried, but modern fish processing plants will use commercial dryers.

Dried maws are ground up, washed, and sterilized before being "cut" by weak acids such as sulfurous acid to disrupt the structure of the collagen molecules so as to generate the correct balance and orientation of positively and negatively charged sites that are responsible for their functionality.

Isinglass is a very pure form of collagen, the selfsame protein that you and I have in our skins (though I don't advocate pumicing over your freshly fermented ale in search of a cheaper, home-grown clarifier). The reason isinglass works rather better than collagen from animal hides can be traced to subtleties in its structure. These capabilities were probably first noticed when people stored beverages in bladders as receptacles.

Rather less widely used, but still an integral part of the process of the world's third biggest Brewer, are wood chips. Over the years these have been mostly derived from well-seasoned beech; individually, they are a few inches wide and as much as a foot long. They therefore present a very ample surface area upon which insoluble materials can stick, including the yeast that is maturing the product.

FILTRATION

After a minimum period of typically three days in this "cold conditioning," the beer is generally filtered. Diverse types of filter are available, perhaps the most common being the plate-and-frame filter, which consists of a series of plates in sequence, over each of which a cloth is hung. The beer is mixed with a filter aid consisting of porous particles that both trap particles and prevent the system from clogging. Two major kinds of filter aid are in regular use: kieselguhr (diatomaceous earth) and perlite. The former comprises fossils or skeletons of primitive organisms called diatoms (figure 10.1). These can be mined and classified to provide grades that differ in their permeability characteristics. Particles of kieselguhr contain pores into which other particles (such as those found in beer) can pass, depending on their size. Unfortunately, there are health concerns associated with kieselguhr, inhalation of its dust adversely affecting the respiratory tract. Pneumatic handling systems are routinely employed to avoid such aerosols.

Perlites are derived from volcanic glasses crushed to form microscopic flat particles. They are better to handle than kieselguhr, but may not be as efficient filter aids.

FIGURE 10.1 Diatomaceous earth (kieselguhr) (courtesy of World Minerals)

BOX 10.2 FILTER AIDS

Kieselguhr, or diatomaceous earth, comprises silica-based shells of ancient unicellular aquatic microscopic plants called diatoms (figure 10.1). Its heat resistance means that it can be used as an insulator, but its abrasiveness means that it has also formed a component of toothpaste and metal polishes. Apart from being widely used as a filter aid to clarify syrups as well as alcoholic beverages, it is used as a filling material in paper, paints, ceramics, soap, and detergents. Oh, and Alfred Nobel found that it is a great absorbent of nitroglycerine in the manufacture of dynamite.

Huge beds of kieselguhr, between 40 and 50 feet deep, are found in California and Virginia, but also in parts of Germany and the United Kingdom, Aberdeenshire in the latter case. The microscopic appearance of kieselguhr from different localities differs considerably. The deposits contain varying amounts of organic matter together with sand, clay, and iron oxide, and the raw material is first incinerated (calcined) to destroy organic matter. The successive process stages in rendering bags of kieselguhr for the brewer's use are mining, crushing, drying, calcining, cooling, air classification, and packaging.

Perlite is a naturally occurring siliceous rock that when heated expands from four to twenty times its original volume. When heated to above 871°C (1600°F), the stuff pops like popcorn to produce many small bubbles, and so perlite is very light and white.

There are many uses for perlite. Its insulating properties and lightness render it valuable as an insulator in masonry and cryogenic vessels. It is used as an aggregate in cement and plasters and for under-floor insulation, chimney linings, paint texturing, gypsum boards, ceiling tiles, and roof insulation boards.

In the world of horticulture, perlite is used as a component of soilless growing mixes, allowing aeration and moisture retention. It is also used as a carrier for fertilizer, herbicides, and pesticides and for pelletizing seed.

We know perlite as a filter agent, and it is also used in this way for cleaning up pharmaceuticals, chemicals, and water. Like kieselguhr, it can also be used as an abrasive.

Filtration starts when a pre-coat of filter aid is applied to the filter by cycling a slurry of filter aid through the plates. This pre-coat is generally of quite a coarse grade, whereas the "body feed" that is dosed into the beer during the filtration proper tends to be a finer grade. It is selected according to the particles within the beer that need to be removed. If a beer contains a lot of yeast but relatively few small particles, a relatively coarse grade is best. If the converse applies, then a fine grade with smaller pores will be used.

The principles of beer filtration are similar to those we encountered when considering lautering (chapter 8). Long filtration runs depend on the conservative application of pressure and are easier to achieve if factors such as viscosity are low. As lower temperatures substantially increase viscosity and as beer should be filtered at as near 0°C (32°F) as possible, it is particularly beneficial if substances like β-glucan are removed prior to this stage. Filtration can proceed until the filter is chock-full of solids, either insolubles removed from the beer or the filter aid itself. For this reason, low solids in the beer and avoidance of excessive levels of filter aid are desirable, lengthening the "filter runs" before the device needs opening up, stripping down, and reestablishing.

STABILIZATION

Apart from filtration, various other treatments may be applied to beer downstream, all with the aim of enhancing the shelf life of the product. There are three principal ways in which beer could deteriorate with time: by staling, by throwing a haze, and by becoming infected. The last of these will be covered in the next section.

As we have seen in chapter 4, the flavor of beer changes in various ways in the package. The most significant of these changes are due to oxidation. It is now generally accepted that oxidation reactions can take place throughout the brewing process and that the tendency to stale can be built into a beer long before it is packaged and dispatched to trade. However, no Brewer would argue with the fact that the oxygen level in the beer as it is filled into its container should be as low as possible. The freshly filtered beer, which is called "bright beer," should have an oxygen content below 0.1 ppm, and many brewers will insist upon substantially lower levels than this. In part this will be achieved by running the beer from the filter into a tank that has been equilibrated in carbon dioxide or even nitrogen. The flow of beer into the vessel will be gentle. And if, once the vessel is full, the oxygen content of the beer exceeds specification, the vessel will be purged with carbon dioxide or nitrogen to drive off the surplus oxygen. Some Brewers (though rarely in the United States) will add antioxidants at this point, such as sulfur dioxide or ascorbic acid (vitamin C), but they are seldom especially useful at this stage.

Brewing scientists have got a long way to go before they will have fully understood the very complex area of beer oxidation. They understand much more about colloidal instability, which is the tendency of beer to throw a haze. As a result, much more robust treatments are available to ensure that beer does not go cloudy within its shelf life.

A haze in beer can be due to various materials, but principally it is due to the cross-linking of certain proteins and certain tannins (polyphenols) in the product. Therefore, if one or both of these materials is removed, the shelf life is extended.

As we have already seen in chapter 8, the brewhouse operations are in part designed to precipitate out protein-tannin complexes. Thus, if these operations are performed efficiently, much of the job of stabilization is achieved. Good, vigorous, "rolling" boils, for instance, will ensure precipitation. Before that, avoidance of the last runnings in the lautering operation will prevent excessive levels of tannin from entering the wort.

We have seen that cold conditioning also has a major role to play, by chilling out protein-polyphenol complexes, enabling them to be taken out on the filter. Control over oxygen and oxidation is just as important for colloidal stability as for flavor stability, because it is particularly the oxidized polyphenols that tend to cross-link with proteins.

For really long shelf lives, though, and certainly if the beer is being shipped to extremes of climate, additional stabilization treatments will be necessary.

In the 1950s, it was shown that nylon could efficiently remove polyphenols from beer. Nylon has rather more stylish applications in society these days, leaving an altogether more efficient if less glamorous material with the job of taking tannins out of beer: polyvinylpolypyrrolidone, which happily is usually abbreviated to PVPP. This can either be dosed into tanks as a solid prior to filtration or impregnated into filter sheets. After use, it can be regenerated by treatment with caustic soda.

Ironically, one of the foremost treatments used to eliminate haze-forming proteins from beer is to *add* more tannin in the form of tannic acid, which is extracted from gall nuts. Although the tannic acid boosts polyphenol levels, this is not a concern, because the proteins that they are able to react with will be removed in the brewery. Indeed, there is a school of thought that better beers contain higher levels of polyphenol, because these molecules contribute to body and also protect against staling through their role as antioxidants. Tannic acid is added at the cold conditioning stage.

There is an increasing usage of silica hydrogels and xerogels to remove haze-forming proteins from beer. These are matrices produced from sand, but in forms that have porous structures able to adsorb macromolecules such as proteins. A range of these products is available, varying in their ability to take up proteins of different sizes. Most importantly, it is claimed that use of these materials does not eliminate the class of proteins that contribute the foam to beer.

A third opportunity to remove haze-forming protein is to add a protein-degrading enzyme to the beer. Most commonly, brewers will use papain, which is derived from the papaya and is the same enzyme that is used in meat tenderizer, but it is known that foam suffers as a result. More recently an enzyme, called prolyl endopeptidase, has come into the market that specifically attacks haze-forming proteins.

BOX 10.3 STABILIZERS

Brewers are comfortable with using those stabilizers that do not get into the product in a soluble form, which of course means silica hydrogels (or xerogels) and polyvinylpoly-pyrrolidone (PVPP).

The silica-based products have their origins in sand, and rather pure sand at that. It is first converted into a soluble form by the action of alkali. Thereafter there is a controlled aggregation of sodium silicate And the washed aggregated particles are processed by techniques including micronization, drying, milling, and classifying to yield the desired balance of particle sizes and pore sizes, the range of which is taken advantage of by brewers (and others) to remove colloidal particles of various types from their products.

PVPP is produced by a technique called "popcorn polymerization." The monomer vinylpyrrolidone is heated with strong caustic and then cooled, in which phase the polymerization takes place. Subsequently there are slurrying, filtering, hydrolysis (using phosphoric acid), washing, reslurrying, and drying stages, in which any residual monomer and water are removed.

Both the silica and PVPP products have been rigorously shown to have impeccable safety credentials. What do you want to remove, proteins or polyphenols? It's your choice, for I have seen no good evidence to suggest that silica hydrogels jeopardize foam by removing protein, or that PVPP has any adverse impact on flavor or flavor stability by getting rid of polyphenols. Indeed, you can now acquire products that incorporate both agents in a single preparation.

To reinforce beer foam, particularly to help its resistance to the damaging effects of oils and fats (see chapter 4), some Brewers add propylene glycol alginate (PGA) to their beer. Like any material used in the brewing industry, PGA has been rigorously evaluated for its wholesomeness; like the Irish moss used in the brewhouse, it is derived from seaweed. (The reader will be struck by the natural origins of the materials used in beer, not only the major raw materials but also processing aids. Apart from PGA and Irish moss, we have isinglass finings, from fish; kieselguhr, which is skeletons of diatoms; tannic acid, from gall nuts; and beech wood chips.)

REMOVING MICROORGANISMS

Although beer is relatively resistant to spoilage (see chapter 4), it is by no means entirely incapable of supporting the growth of microorganisms. For this reason, most beers are treated to eliminate any residual brewing yeast or wild yeasts and bacteria that might infect it before or during packaging. This can be achieved in one of two ways: pasteurization or filtration.

PASTEURIZATION. This can take one of two forms in the brewery: flash pasteurization for beer pre-package, typically on its way to filling in to kegs or heat-sensitive plastic bottles, and tunnel pasteurization for beer in can or bottle. The principle in either case, of course, is that heat kills microorganisms. The higher the temperature, the more rapidly are microorganisms destroyed.

In flash pasteurization, the beer flows through a heat exchanger (essentially like a wort cooler acting in reverse—see chapter 8), which raises the temperature typically to 72°C (162°F). Residence times of between 30 and 60 seconds at this temperature are sufficient to kill off virtually all microbes. Ideally, there won't be many of these to remove; good Brewers will ensure low loading of microorganisms by attention to hygiene throughout the process and ensuring that the previous filtration operation is efficient. The configuration of the flash pasteurizer is such that heat from the beer leaving the device is used to warm the entering beer. It is essential that the oxygen level of the beer be as low as possible before pasteurization, because when temperatures are high, oxygen is "cooked" into the product, giving unpleasant flavors.

Tunnel pasteurizers comprise large heated chambers through which filled and sealed cans or glass bottles are conveyed over a period of minutes, as opposed to the seconds employed in a flash pasteurizer. Accordingly, temperatures in a tunnel pasteurizer are lower, typically 60°C (140°F) for a residence time of 10–20 minutes.

STERILE FILTRATION. An increasingly popular mechanism for removing microorganisms is to filter them out by passing the beer through a fine mesh filter. The rationale for selecting this procedure rather than pasteurization is as much for marketing reasons as for any technical advantage it presents; many brands of beer these days are being sold on a claim of not being heat-treated, and therefore free from any "cooking." In fact, provided the oxygen level is very low, modest heating of beer does not have any impact on the flavor of most beers, although those products with relatively subtle, lighter flavor will obviously display "cooked" notes more readily than will beers with a more complex flavor character.

GAS CONTROL

Apart from stabilization downstream, final adjustment will be made to the level of gases in the beer. As we have seen, it is important that the oxygen level in the bright beer be as low as possible. Unfortunately, whenever beer is moved around and processed in a brewery there is always the risk of oxygen pickup. For example, oxygen can enter through leaky pumps. A check on oxygen content will be made once the bright beer tank is filled, and if the level is above specification, oxygen will have to be removed. This is achieved by purging the tank with an inert gas, usually nitrogen, from a sinter in the base of the vessel. It is not a desirable practice, because whenever a purging process takes place there is a foaming on the beer. The foam sticks to the side of the tank and dries, the resulting flakes falling into the beer to form unsightly bits.

The level of carbon dioxide in a beer may either need to be increased or decreased. The majority of beers contain between 2 and 3 volumes of CO_2, whereas most brewery fermentations generate "naturally" no more than 1.2–1.7 volumes of the gas. The simplest and most usual procedure by which CO_2 is introduced is by injection as a flow of bubbles as beer is transferred from the filter to the bright beer tank. If the CO_2 content needs to be dropped, this is a more formidable challenge. It may be necessary for beers that are supposed to have a relatively low carbonation (beers such as the nitrokegs or draft-beers-in-can discussed earlier), and, as for oxygen, this can be achieved by purging. However, concerns about "bit" production have stimulated the development of gentle membrane-based systems for gas control. Beer is flowed past membranes that are water-hating and therefore don't "wet out." Gases, but not liquids, will pass freely across such membranes, the rate of flux being proportional to the concentration of each individual gas and dependent also on the rate at which the beer flows past the membrane. If the CO_2 content on the other side of the membrane is lower than that in the beer, the level of carbonation in the beer will decrease. If the CO_2 content on the other side of the membrane is higher than that in the beer, then the beer will become more highly carbonated. Gases behave independently, so the membranes can be used simultaneously to remove CO_2 from a beer and also to remove any oxygen from it, provided the levels of both gases is lower on the other side of the membrane. This technique is also an excellent opportunity to introduce nitrogen into beer, a gas which we have seen (in chapter 4) has tremendous benefits for beer foam.

PACKAGING

The final process stage, prior to warehousing of the beer, is to put it into the intended package. In the United States the balance of packaging is some 9% on draft, 51% in cans, 38% in nonreturnable glass bottles, and the remainder in glass bottles returned to the Brewer for washing and refilling. As we saw in chapter 1, Ireland and the United Kingdom sell a large proportion of their beer on draft dispense. In most other countries, the favored package is the bottle, usually the kind that is returned to the brewery for washing and reuse. In France and Italy, though, it is in nonreturnable glass that much of the beer is retailed, while Sweden is the country that sells the highest proportion of beer in can. It is only relatively recently that pressures from the European Community came to bear in Denmark, obliging them to release beer in cans. Back in North America, a transparent difference between "small pack" beer in the United States and Canada is the heavy preponderance of nonreturnable glass in the United States, whereas in Canada the bulk of bottled beer is in returnable glass.

Time was when all beer was on draft, in that beer was purchased from the alehouse or even the brewery in earthenware and pewter jugs and diverse other receptacles. In the United Kingdom it was the removal of a tax on glass that stimulated the bottling of beer as the twentieth century dawned.

BOX 10.4 PACKAGING MATERIALS

Roughly half of the cost of a pallet of filled beer bottles is accounted for by the cost of bottles and trays. Beer represents only 25% of the value. For canned beer, some 70% of the batch cost is the can—which of course embraces the label and closure.

BOTTLES

Although there is no evidence that the very first Sumerian brewers of beer had the gumption to bottle their wares, glass produced by volcanic action was known all those thousands of years ago. I guess those folk didn't know how to fashion it into containers, though it does seem that the Phoenician sailors at least sussed out how to make glass.

The starting point is silica in the form of sand. Some 72% of this is mixed with 14% soda ash, 12% lime, 2% alumina-based stabilizer and colorants: iron sulfate in the case of amber glass and iron chromite for the green stuff.

The beauty of glass is that it is impermeable (e.g., to gasses), it doesn't corrode, and it is chemically inert.

The design of the common or garden beer bottle is a compromise involving stresses and strains. The most resistant shape for a glass bottle is a perfect sphere, but to the best of my knowledge no brewer turns out product looking like a soccer ball. Second best is a cylinder with semispherical ends. And so with the beer bottle we have the next best thing. Of course there is a diversity of beer bottle shapes, but seldom do they stray terribly far from the cylindrical body with curvy base. Subtle and stylish differences in the neck region, however, will lead to a greater or lesser difference in the robustness of the container.

The stress in a beer bottle increases in proportion to the internal pressure (read "gas content") and the diameter of the bottle. It decreases as the thickness of the glass is raised.

(continued)

BOX 10.4 CONTINUED

Beer bottles are treated to further increase their strength, although these appli-cations are of no use for returnable bottles because the caustic and the materials released from digested labels screw up the bottle surface. Although such bottles do tend to be treated with water-soluble sodium stearate–based agents, returnable glass is treated with two types of permanent application; the newly made hot bottle is exposed to a vaporized metallic formulation ("hot end treatment"), which forms a nice base for the application after cooling (though with the glass still at 120°C) of polyethylene or perhaps oleic acid ("cold end treatment"). This latter treatment lubricates the glass so that the bottles don't scuff when they bang against one another and the rails on the packaging line.

The first stage in making a bottle, of course, is the weighing of the various compo-nents of the glass, followed by highly efficient mixing to ensure homogeneity. It is com-mon (for environmental reasons) to add to the mix some "cullet," which is recycled glass. Naturally it needs to be the appropriate color and free of junk, crown corks for instance.

Next stop is the furnace, where basic glass is made and the molten contents may reach 1500°C (2732°F). From here it's through a conditioning stage (1150°C/ 2102°F) to the place where the bottle is fabricated. Stage one here is the deliciously named "gob feeder," where the basic quantity of glass of a shape that will subsequently be moldable is delivered by a plunger to a set of shears that slice off the appropriate quantity for one bottle. The gob feeder discharges these units at around 1100°C (2012°F) to the molding devices, which "blow" the bottle into its final dimensions and set it on its way at 510°C (950°F). These huge temperature differences introduce stresses, so the newly formed bottle heads off to an annealing stage where the temperature is ramped up to 600°C (1112°F) and held briefly to relieve tension.

CROWN CORKS

The crown closure dates back to William Painter's patented device of 1892, with its natu-ral cork lining. Subsequent developments have largely attended to the lining materials. Devastation of forests in search of cork for this and other uses have led to plasticized polyvinyl chloride (PVC) as the norm since the mid-1950s, with the more recent addi-tion of oxygen-scavenging inlays as brewers have sought protection against the insidi-ous creep of O_2 through the crown cork–glass seal. The "twist off" crown cork was first introduced in 1966 (am I alone in having such precious flesh that I need a handkerchief to make the removal of these anything less than painful?).

The making of crowns involves the coating and decorating (lithographing) of metal sheets, followed by the mechanical punching out and lining of the crowns. A low metal-loid steel is given a tin or chromium-based anti-rust coating. Just as for glass, there is an annealing process to relieve stresses, this time in the base steel, and to soften the metal in advance of punching. The coated metal is afforded a thin coat of lubricant (e.g., butyl stearate) to aid subsequent handling. Particularly for the twist-off crowns, lubricant needs to be mixed with the PVC formulation.

LABELS

A number of factors enter into the selection of paper for application to beer bottles, whether the main or secondary labels or the neck foil. Apart from cost, criteria include smoothness, curl characteristics, and water penetration/wet strength.

For foil labels, repeated rolling of pure aluminum to a thickness of 0.00014 cm precedes lamination by water-resistant glue onto a paper backing of sufficient robustness to withstand printing and handling.

The main labels may be metallized or varnished in the interests of surface appearance. This is also relevant to the behavior of the paper during label manufacture, which basically comprises printing, embossing, and cutting. Issues include the tendency of labels to stick together at the edges after cutting.

A quick mention of adhesives: they may be casein- or resin-based. The latter have increased ice-resistance. (I'll never forget the time I was a guest at the home of a senior executive of a mega-brewing company and took a bottle from an ice bucket only to see it separate from its label in my hand to propel the bottle as a missile in the direction of mine host.)

CANS

Cans as we know them these days were first conceived in 1962, with the lift-tab, easy-to-open aluminum end that followed on from the development in 1958 of the two-piece can (with the base formed as part of the body mold rather than applied as a separate piece, just as for the top end). The lift-tab quickly became the ring pull, but nowadays the norm is the stay-on tab, which first saw the light of day in 1975. Most cans these days are aluminum-based, but the processes for making these and tinplate cans are similar.

Metal in the form of a coil is unwound and lubricated before feeding into a device that forms the metal into shallow cups. These go into an ironing press, where the can wall and bottom profile are formed. The latter is critical in relation to strength demands and resistance to pasteurization. The top end is trimmed and lubricant removed by spraying before the can is decorated on rotary equipment followed by oven hardening. The final stage is die necking, whereby the neck is narrowed to the final dimension. Recent years have been marked by a narrowing of the neck size in the interests of metal conservation.

The ends are produced from a coated flat stock that is baked, cut into strips, and fed into a press to stamp out the end with all its features, for instance nondetachable pull. From there it goes to a curling unit to turn the edge inward and apply a sealant.

(continued)

BOX 10.4 CONTINUED

PLASTIC BOTTLES

Connoisseurs reject plastic bottles as being barbarous and against tradition. It seems that mainstream customers will need to get used to the concept but see the functional benefits, not least, one assumes, the ability to swig with neither spillage nor threat down the football stadium. Housewives like the lightness and functionality of this medium. Remember that plastic weighs in at least five times lighter than glass.

For many, glass is much more in keeping with beer as a package container; indeed, cans are just as controversial as plastic. Rightly or wrongly, customers perceive cans as delivering a metallic taste and they are regarded as being worse for the environment than glass. Cans don't afford the same comfort factor in physical drinking (for those who can't be bothered to pour the beer into a glass). One of the best arguments for cans, of course, is that you can dangle them over the side of the boat when fishing to keep the beer nice and cold.

Regarding plastic, though, it has been reported that consumers like the resealable top, the recyclability, and the nonslip surface, and, frankly, they are prepared to say that plastic beer bottles these days *look* like glass ones. In an age when it is predicted that the international beer market will soon employ over 300 billion containers each year, which laid end on end would encompass our planet more than a thousand times, it is worth even the most ardent traditionalist's pausing in consideration of what plastics technology might offer the brewer.

The early days of beer in plastic were not auspicious—who didn't take big plastic bottles of cheap beer and cider to parties in the early '70s and grumble about the pourability and quality? (Well, actually, I didn't: I took cheap draft port, but it gives me a headache just to think of it.) Those were the days of simple PET (polyethylene terephthalate), with its high gas permeability and attendant loss of CO_2 and pickup of O_2.

Plastics technology has advanced to the point that now we have multilayer bottles with one or more gas-barrier or absorbing layers. Alternatively, such bottles may have external or internal coatings to reduce gas diffusion. There are preferable alternatives to PET, such as nanocomposites and polyethylene naphthalate (PEN). PEN is very expensive and therefore only used for returnables.

External coatings and nanocomposites inserted into PET are made by generating a plasma using vacuum, the alternative being to spray on coatings. The bottles themselves are made by blowing technology—they may even be "erected" in the packaging plant, immediately before filling.

The first trials on putting beer into cans took place in post-Prohibition New Jersey, when the Krueger Brewing Company of Newark first sold canned beer in January 1935. It was an immediate success, being a packaging medium that was light, nonbreakable, and furthermore protected beer absolutely from the adverse influences of light. Just one year later, the Welsh Brewer, Felinfoel, emulated Krueger, taking advantage of the presence of the can manufacturing capabilities of the steel industry nearby. Canned beer now accounts for a quarter of the beer production in the United Kingdom, a proportion that has increased substantially in recent years following the development of the "widget" and the shift toward drinking at home. Enormous improvements in the barrier and feel properties of plastics, which mean that they are much less likely these days to permit oxygen ingress and also have a more prestigious appearance, means that there is likely to be an ongoing sizable swing to beer in this packaging matrix. This will greatly enhance retail opportunities as well as light-weighting beer for shipment.

The traditional package for beer in the United Kingdom, of course, is the cask, originally made from wood by coopers but increasingly composed of aluminum or stainless steel. There is still a healthy market for cask-conditioned ale in the United Kingdom, beer that is not pasteurized and that retains yeast within it to naturally carbonate the product. The yeast is settled out from the beer using isinglass finings, and it is essential that the beer be handled carefully to avoid disturbing the sediment, rendering the beer cloudy. And if it is disturbed, the sediment should be capable of resettling, perhaps several times.

Although the remarkable growth of microbrewers in the United States has reintroduced ale increasingly into the consciousness of the U.S. drinker, such products have not always achieved total acclaim among American drinkers. Certainly, the men of the U.S. Air Force stationed in East Anglia, England, in World War Two didn't care to have cloudy beer delivered to them as they returned from missions, and a shortage of glass meant that bottled beer was out of the question. The quandary prompted Air Force General Curtis LeMay to approach a nearby brewer, Greens of Luton, to see how they could overcome the problem. Over $150,000 was spent on developing the process of putting into metal barrels beer that was carbonated and sediment-free. "Keg" beer was born.

FILLING BOTTLES. Glass bottles used for holding beer come in diverse shapes and sizes. The glass may be brown or black, green or clear (which is usually referred to as flint glass). Marketing people are increasingly obsessed with beer being packaged in any color of glass other than brown. They should listen to their technical colleagues: as we saw in chapter 4, unless precautions are taken, beer develops a pronounced skunky character within seconds of exposure to light. Brown (or black) glass minimizes the access of light to beer, whereas green or flint glass provides no protection whatsoever.

Bottles entering the brewery's packaging hall are first washed, irrespective of whether they are one-trip or returnable. For the former, they will receive simply a water wash, as the supplier will have been required to make sure they arrive at the plant in a clean state. Returnable bottles, after they have been automatically removed from their crate and delivered to conveyors, need a much more robust clean and sterilization, inside and out, involving soaking and jetting with hot caustic detergent and thorough rinsing with water. Old labels will be soaked off in the process. The cleaned and sterilized bottles pass an empty bottle inspector (EBI), a light-based detection system that spots any foreign body lurking in the bottle. Now they're on their way to the filler.

The beer coming from the bright beer tanks after filtration is transferred to a bowl at the heart of the filling machine. Bottle fillers are machines based on a rotary carousel principle. They have a series of filling heads: the more heads, the greater the capacity of the filler. Modern bottling halls will be capable of filling in excess of 1,200 bottles per minute. If you go into the bottling hall, you will see these mighty beasts whirling round, with empty bottles chinking their way toward them and full ones whizzing away from them.

The bottles enter on a conveyor and each in turn is raised into position beneath the next vacant filler head, each of which comprises a filler tube. An airtight seal is made, and, in modern fillers, a specific air evacuation stage starts the filling sequence (we have already seen how damaging oxygen is to beer quality). The bottle is counterpressured with carbon dioxide before a valve is opened to allow the beer to flow into the bottle by gravity from the bowl. The machine will have been adjusted so that the valve is open long enough to allow the correct volume of beer to be introduced. Once filled, the "top" pressure on the bottle is relieved, and the bottle is released from its filling head. It passes rapidly to the machine that will crimp on the crown cork, but en route the bottle will have been either tapped or its contents "jetted" with a minuscule amount of sterile water in order to fob the contents of the bottle (i.e., make them foam up) and drive off any air from the space between the surface of the beer and the neck (the "headspace").

Next stop is the tunnel pasteurizer (see above) if the beer is to be pasteurized after filling—although, as we have seen, more and more beer is sterile-filtered and packaged into already sterilized bottles. In the latter case, the filler and capper will likely be enclosed in a sterile room to which only necessary personnel are allowed access.

The bottles now pass via a scanner, which checks that they are filled to the correct level (if not, they are rejected), to the labeler, where labels are rolled on to the bottles and then perhaps to a device that will apply foil over the cap. Other specialist equipment may involve jetting on a packaging date, "best before" date, or "born on" date. Finally, the bottles are picked up by a machine that places them carefully into a crate, or box, or whichever is the secondary package in which they will be transferred to the customer. Perhaps they will go straight from this operation onto a truck or railcar for shipping, but more frequently they will be stored carefully in a warehouse prior to release.

CANNING. Putting beer into cans has much in common with bottling. It is the container, of course, that is very different—and definitely one-trip.

Cans may be of aluminum or stainless steel, which will have an internal lacquer to protect the beer from the metal surface and vice versa. They arrive in the canning hall on vast trays, all preprinted and instantly recognizable. They are inverted, washed, and sprayed, prior to filling in a manner very similar to the bottles. Once filled, the lid is fitted to the can basically by folding the two pieces of metal together to make a secure seam past which neither beer nor gas can pass. (To get an idea for this, bend the fingers on both of your hands toward the palms, then put the right hand palm downward on top of the left hand palm up before sliding the right hand toward the right until the ends of the fingers on the right hand are tight underneath those on the left hand. Squeeze the fingers on both hands toward your palms: the tight fit you have created is exactly analogous to the seal between a can and its lid.)

KEGGING. Kegs are manufactured from either aluminum or stainless steel. They are containers generally of 1 hectoliter or less, containing a central "spear" (tube) through which the keg is washed, filled, and emptied in the bar. Kegs, of course, are multitrip devices. On return to the brewery from an "outlet" they are washed externally before transfer to the multihead machine, in which successive heads are responsible for their washing, sterilizing, and filling. Generally they will be inverted as this takes place. The cleaning involves high-pressure spraying of the entire internal surface of the vessel with water at approximately 70°C (158°F). After about ten seconds, the keg passes to the steaming stage, the temperature reaching 105°C (221°F) over a period of perhaps half a minute. Then the keg goes to the filling head, where a brief purge with carbon dioxide precedes the introduction of beer, which may take a couple of minutes. The discharged keg is weighed to ensure that it contains the correct quantity of beer and is labeled and palleted before warehousing.

Right to the last process stage, then, with the weighing of the kegs, the Brewer is conscientiously ensuring that the product is precisely right for the consumer. As we have seen in chapters 5 through 10, the Maltster and Brewer operate processes that are carefully controlled to ensure consistency. To help them achieve this, they need procedures for measuring the raw materials and the various streams and for analyzing the finished product in order that they can be satisfied that everything is in order. So far, I have mentioned the sorts of measures that are taken to monitor the raw materials and the process. Now, in chapter 11, we will find out how the Brewer analyses the beer itself.

MEASURE FOR MEASURE

HOW BEER IS ANALYZED

A former colleague of mine used to talk of his boyhood, and of his father coming home from the pub.

"That was a good pint tonight," the father would announce, doubtless licking his lips. The implication was that, some evenings, it wouldn't be a good pint.

These days, the production of beer is marked by strong quality control. Indeed, breweries are as aware as any industry of the merits of applying principles of quality assurance, in respect of an ethos of "right first time" and backed up by adherence to standards such as ISO 9002.

As we have seen, malting and brewing are not simple processes. They are marked by a complex blend of vegetative and mechanical stages, at any of which there is plenty of opportunity for things to go wrong. That this is seldom the case is testimony to the skill of the Maltster and the Brewer—and to the availability of robust analytical methodology.

THE ANALYSIS OF BEER

A Brewer would not succeed if her measurements were made on finished beer alone. Throughout this book, I have drawn attention to the sorts of specifications that are made on raw materials and in individual process stages. The establishment of specifications demands the availability of methodology to make the necessary measurements. Wherever possible, Brewers seek to install sensors to enable them to make their measurements automatically, together with associated control systems that respond to the values measured and, if values are out of specification, adjust a relevant parameter in order to push the process back on track. For example, temperature is readily measured remotely during fermentation and, if it rises, can be automatically lowered by triggering the circulation of coolant through the jackets of the fermenters.

Temperature is one of the fundamental measurements that need to be made throughout the malting and brewing processes in order that they can be controlled. Others include weights, rates of liquid flow, pressure, and fill heights in vessels. Table 11.1 lists the other parameters that are routinely checked in a brewery to confirm that the process is progressing according to plan at all stages.

This chapter concentrates on the analysis of the finished beer itself. The techniques applied are used to confirm that a batch of beer is acceptable for packaging and for subsequent release into the trade. Equally, some of the methods will be used in the trade to confirm that the product is in good condition.

They can also be applied to assess a competitor's beers, to see what "tricks" they are employing and to try to unravel some of the procedures that they are

TABLE 11.1. MINIMUM ANALYSES THAT SHOULD BE MADE AND RESPONDED TO FOR THE BREWING PROCESS TO BE KEPT UNDER CONTROL

Parameter	Methodology
Absence of taints in liquor supply	Taste it daily
Specific gravity of wort collected in brewhouse, when fermenter is filled and during fermentation	Hydrometer or "vibrating" U-tube instrumentation
Dissolved oxygen in wort pre-yeast dosing	Oxygen sensor
Amount of yeast "pitched"	Hemocytometer, sensors based on light scatter or capacitance
Vicinal diketones in freshly fermented beer	Spectrophotometry or gas chromatography
Alcohol content of beer for declaring duty and controlling dilution of high-gravity brews	Various, including distillation, gas chromatography, or near-infrared spectroscopy
Gases (CO_2, O_2, N_2) in bright beer	Specific gas sensors
Clarity of bright beer	Hazemeter
Color of bright beer	Spectrophotometer, Tintometer, tristimulus colorimeter
Bitterness of bright beer	Spectrophotometer, high performance liquid chromatography
Parameters during packaging (alcohol, gases, color, contents, integrity of seams between can and lid)	Contents by weighing Physical strip-down and visual examination for seam checks
Caustic strength of CIP detergent	Titrate
Flavor acceptability	Taste contents of representative samples from all packaging runs

using to make a particular beer. Having said this, and despite the fierce competition that exists between Brewers, they do share a spirit of cooperation in establishing the methodology that will be used to measure their products. The driving forces for this are severalfold. For instance, Brewers must clearly use methods for measuring alcohol that enable direct comparison of the strength of their various products for duty declaration purposes and for letting the consumer know how strong a given product is. Secondly, there is much cross-brewing of beers: Brewers may well franchise-brew the products of a competitor. There is self-evidently a need for a common "language" to describe the attributes of a beer.

For these reasons, Brewers come together through various forums. In 1886, the Laboratory Club was set up at a meeting in, of all places, a coffeehouse in Fitzroy Square, London, to act as a meeting point for British Brewers to enable them to share experiences. It developed into the Institute of Brewing (IOB), now known as the Institute of Brewing & Distilling (IBD), which now serves this purpose on an international stage. Among its roles is the publication and evolution of a set of standardized methods. Relevant methods are debated in committee before being written up in a standardized format that is clearly understandable and in a form that any laboratory should be able to faithfully follow. The method and samples for measurement are circulated to a wide range of laboratories, each of which produces a set of data. This is collated and analyzed statistically by the Committee, which is able to assign values that indicate how consistent the results are when a method is applied by the same analyst in a single location or by analysts in different locations. Only if these values indicate good consistency and agreement will any confidence be placed in a method's ability to give reliable and reproducible values that can be used not only for process control but also as a basis for transactions.

Similar activities occur within the European Brewery Convention (EBC) and the American Society of Brewing Chemists (ASBC). There are clear differences between the various sets of methods, but lots of similarities, too, and measures have been taken to harmonize at least the methods of the IBD and EBC. Pressures to prevent this are largely founded in history, in that the IBD methods relate more closely to technology employed in the British Isles (and some but not all of Britain's old colonies!), whereas the EBC methods relate to continental brewing techniques. As brewing companies become more international and individual brands break down national barriers, it is the state of origin of a beer that tends to dictate how it will be analyzed.

The methods can be classified in several ways. Perhaps the most useful for our purposes is into chemical analysis, microbiological analysis, and organoleptic analysis.

CHEMICAL ANALYSIS

ALCOHOL. Perhaps the most critical measure made on beer is its content of alcohol. In many countries (although the United States is not one of them), tax (duty) is levied on the basis of alcohol content. In the United Kingdom, for instance, the rate of duty collection is in proportion to how much alcohol there is. And you wondered why the tendency is toward lower alcohol contents in the United Kingdom?

A Brewer in the United Kingdom needs to be able to declare the alcohol at least to within an accuracy of 0.1%. The methodology employed can vary: HM Revenue & Customs stipulates only that a method should be used that can be proven to give sufficiently precise results. Most commonly alcohol will be measured by gas chromatography, but other methods may include distillation and specific alcohol sensors.

Allied to the declaration of alcohol, the Brewer must also identify for Customs purposes (and to satisfy weights and measures legislation) the volume of beer that is being produced for sale. This is generally established on a container-by-container basis by weighing the vessel, be it a keg, can, or bottle. Application of statistical distribution analysis indicates whether the inevitable spread of weights across a population of containers is within acceptable limits.

Accurate measurement of alcohol is also necessary to control the strength of beer produced by high-gravity fermentation techniques. As we saw in chapter 9, it is common practice for fermentation to be performed in a concentrated state, with the beer diluted just prior to packaging. This dilution is controlled on the basis of alcohol content, with deaerated water being added to bring the alcohol content down to that specified for the beer in question. In many breweries this control is carried out in-line. A sensor prior to the dilution point measures the alcohol content continuously and regulates the rate of flow of beer and water at the subsequent mixing point. The alcohol-measuring sensor may be based on one of several principles, one of the most common being near-infrared spectroscopy.

CARBON DIOXIDE. Just as carbon dioxide is produced hand in hand with ethanol in fermentation, so is it a critical parameter to be specified in the finished product. The level of CO_2 will be measured in the bright beer tank, most frequently using an instrument that quantifies CO_2 on the basis of pressure measurement. If the gas level is too low, CO_2 is bubbled in to meet the appropriate specification. If the level is too high, carbonation can be delivered down to specification either by sparging with nitrogen or by the use of hydrophobic gas control membranes (see chapter 10).

ORIGINAL EXTRACT AND RESIDUAL EXTRACT. The term "original extract" is frequently encountered. Allied to the measurement of alcohol, it is an indicator of the strength of a product. If the alcohol content of a beer is known, it is possible to calculate the quantity of fermentable sugar that must have been present in the wort prior to fermentation. This can be added to the real extract (sometimes called the "residual extract," which comprises nonfermented material, primarily dextrins) to obtain a value for the original extract. The real extract is determined as specific gravity by using a hydrometer or pycnometer or, more commonly these days, a gravity meter. These operate on the basis of vibrating a U-tube filled with the beer. The frequency of oscillation relates to how much material is dissolved in the sample. The real extract tells the brewer whether the balance of fermentable to nonfermentable carbohydrate in the wort was correct and whether the fermentability of the wort was too high or too low.

pH. Another indicator of fermentation performance is the pH of the beer. During fermentation, acids such as citric and acetic acid are secreted by yeast and the pH drops. The more vigorous and extensive the fermentation, the lower the pH goes. The pH has a substantial effect on beer quality (see chapter 4), not least by its influence on flavor and its influence in suppressing microbial growth; it is measured using a pH electrode.

COLOR. All beers have their characteristic color, whether it is the paleness of lagers or the intense darkness of a stout. The most frequently used procedure for assessing color is by measurement of the absorbency of light at a wavelength of 430 nm. For the lighter products there is a reasonable correlation between this value and color, but problems may occur with darker beers. The perception of color by the human eye depends on the assessment of absorption at all wavelengths in the visible spectrum. It is no surprise, then, that a panel of expert judges could tell apart beers displaying identical absorption of light at 430 nm but with small yet significant differences in hue. The modern standard for color measurement employed in many industries is based on tristimulus values, which basically describe color in terms of its relative lightness and darkness and its hue. The nearest thing to it is a technique employed by many traditionalists for a great many years, namely the comparison of the color of the beer with that of each of several discs in a device called a Lovibond Tintometer.

CLARITY. Another key visual stimulus in beer is its brightness or clarity. Although there are a few beers in the world that are intended to be turbid to a greater or lesser extent (the delightful Coopers Brewery in Adelaide, South

Australia, presents one such example), for most beers cloudiness is undesirable. Haze is measured in beer by the assessment of light scatter by particles. Traditionally, this was by shining light through the beer and measuring the amount of light scattered at an angle of 90°. The more light scattered, the greater the haze. For most beers, there is good agreement between the amount of light scattered in this way and the perceived clarity of the product—but not for all. Sometimes a beer may contain extremely small particles that are not readily visible by the human eye but that scatter light strongly at 90°. The beer looks bright, but the hazemeter tells a different story. This phenomenon is called "invisible haze" or "pseudohaze." It doesn't present a quality problem in the trade, but it is highly problematic for the brewer, who is forced to make a qualitative judgment as to whether a beer rejected instrumentally is satisfactory for release to trade after all. Nowadays there are hazemeters that read light scatter at 13° rather than at right angles, and these don't pick up invisible haze. Unfortunately, they miss some of the bigger particles, which are detected by 90° scatter. Accordingly, some Brewers measure light scatter at both angles—but all will ultimately look at the beer as the acid test!

DISSOLVED OXYGEN. Even though a beer is bright when freshly packaged, it may develop haze after a greater or lesser period of time in the trade. One of the causes of this could be a high level of oxygen in the package. An even more likely problem if levels of this gas were high would be the onset of staleness in the beer. Brewers, therefore, are rigorous about excluding oxygen from the packaged beer (and, to an increasing extent, they try to exclude oxygen further and further back in the process). Reliable measurement of oxygen is essential, and this is generally carried out using an electrode based on principles of electrochemistry, voltammetry, or polarography. It must be carried out before any pasteurization, for the heating will "cook in" the oxygen.

PREDICTION OF STABILITY. Oxygen is only one factor that will influence the physical breakdown of a beer. The most common building blocks of a beer haze are proteins and polyphenols (tannins). As yet, nobody has proved which of the proteins in beer are particularly prone to throw hazes, and until this is rectified, the only way to test the level of haze-susceptible protein is to "titrate" them. In some quality control laboratories, samples of beer will be dosed with aliquots of either ammonium sulfate or tannic acid. The more of these agents are needed to precipitate out protein and throw a haze, the less haze-forming protein is present. Many Brewers measure the other components of haze, the polyphenols. These can be quantified by measuring the extent of color formation when beer is reacted with iron in alkaline solution. Although this measures total polyphenols and they are not all harmful (for instance, some are likely to be antioxidants), it is a very useful means for checking whether a polyphenol adsorbent such as PVPP has done its job or whether it needs to be regenerated. Most frequently, beer stability is forecast through the use of breakdown tests. Beer may be subjected, for instance, to alternate hot and cold cycles, to try to simulate storage in a more rapid time frame.

BITTERNESS. Most Brewers rely on a method introduced over forty years ago by a famed American brewing scientist, Mort Brenner. It involves extracting the bitter substances from beer with the solvent isooctane and measuring the amount of ultraviolet light that this solution absorbs at 275 nm. The greater the absorbency, the greater the bitterness. There is much debate about the use of high performance liquid chromatography (HPLC) for the measurement of the level of bitterness in beers, and this certainly would enable the quantization of the six different species that contribute to bitterness. Technically, then, a closer measure of the actual bitterness of a beer should be obtainable.

DIACETYL. All responsible Brewers will measure the level of this compound in their beers. As we saw in chapter 8, it is produced in all brewery fermentations, which must be prolonged until such time as the yeast has consumed it. A colorimetric method is available to measure diacetyl, but more frequently it is assessed by gas chromatography. It is important not only to measure free diacetyl, but also its immediate precursor, a substance called acetolactate. If any of the latter is left in the beer, it can break down to release diacetyl in the package, giving a most unpleasant butterscotch character to the beer. Before the gas chromatography, therefore, the beer is warmed to break down any precursor to diacetyl.

OTHER FLAVOR COMPOUNDS. Diacetyl is easily the most frequently analyzed flavor component of beer. Some Brewers will measure others as well, but for all Brewers it is through smelling and tasting the beer that they will make their key assessment of its acceptability and judge whether it can be released to trade. Latterly, trials have been undertaken with so-called artificial noses, sensors that are claimed to be able to mimic the human olfactory system. They are far from ready for the job (doubtless to the satisfaction of brewers everywhere!). Among the volatiles that the brewing quality control lab may be required to measure, by gas chromatography, are dimethyl sulfide and a range of esters and fusel oils. It is most likely that this will be on a survey basis, perhaps monthly, rather than brew by brew.

FOAM STABILITY AND CLING. By measuring the carbon dioxide content, the brewster has an index of whether a beer has sufficient capability to generate a foam. This will not tell her whether the resultant foam will be stable, for which another type of analysis is necessary. This is a difficult task, and there is much debate over the best way to measure foam stability (see box 11.1).

METALS AND OTHER IONS. Several inorganic ions are measured in the brewery, mostly on a survey basis. Iron and copper are very bad news for beer, as they promote oxidation (otherwise, iron would be a useful foam stabilizer). They are measured by atomic absorption spectroscopy, as is calcium. Liquid

---

BOX 11.1 MEASUREMENT OF BEER FOAM STABILITY

It is one of the great truths of analytical science that if there are a lot of methods for measuring something, then none of them can be much good. There are a lot of methods for measuring foam quality—at least twenty!

In the United States, one of the most widely used procedures is the so-called sigma value method. In this test, foam is produced by pouring the beer into a specially designed funnel and the stability of the foam calculated from an equation that compares the amount of beer that has drained from the foam in a period of 3–4 minutes with the amount of beer that is still held in the foam itself. This method, therefore, depends on measuring the rate at which beer drains from foam: the more slowly the beer re-forms as a liquid, the more stable the foam.

Derek Rudin, forty years ago, developed another drainage procedure that employs a long, thin glass tube. A little beer is introduced into the bottom of the tube before carbon dioxide is bubbled through it to convert it all into foam, which rises up the tube. When the top of the foam hits a line marked off on the glass tube, the gas supply is switched off. The foam, of course, starts to collapse, and as it does so, the beer starts to re-form at the bottom of the column. The analyst, armed with a stop watch, measures the rate at which the beer re-forms by timing the rise of the foam-beer interface in terms of the seconds or minutes it takes to pass between two more marks on the glass tube. The longer this period of time, the more stable the foam.

A third device, this one developed by a Dutchman called Walter Klopper in the '70s and called the NIBEM method, works on a different principle. Here, the beer is poured into a glass and a plate with needles on it is lowered into the top of the foam. These needles sense the conductivity of the foam (suffice to say that this enables the needles to differentiate the liquid in the foam from the air above the foam). As the foam collapses, the needles "lose" the conductivity signal and send a message to a motor that lowers the needles until the foam is contacted again. This continues as the foam collapses; clearly, the more rapidly the needles lower, the less stable is the foam. The rate of lowering is flashed up as a digital readout in terms of seconds.

chromatography is used to detect the levels of a range of anions, such as chloride and sulfate.

MICROBIOLOGICAL ANALYSIS

A few years ago, a keen young brewer was inspecting the open square fermenters in an old-fashioned English brewery when he spotted a thick crust caked onto the inside. He scraped a lump off with his hand and marched in to see the head brewer.

"What's that, lad?" said the old man.

"It's dirt from the top of a fermenter," replied the young fellow, proud of his discovery and firmly resolved to clean up the plant.

"Well, put it back," stormed the boss. "Where do you think the character comes from in our beer?"

The story is apocryphal (I think!). It does, however, serve to remind us that, despite the fact that beer is relatively resistant to microbial infection thanks to hops (see chapter 6), ethanol, low pH, high CO_2, and lack of oxygen, there is still plenty of opportunity for organisms to infect the process and the product.

Traditionally, microbiological analysis in breweries consisted of taking samples throughout the process and inoculating them on agar-solidified growth media of various types designed to grow up specific categories of bacteria or "wild yeasts" (any yeast other than the one used to brew the beer in question). When the plates were incubated for 3–7 days, any bugs on them would grow to produce colonies: the more colonies, the greater the contamination. The problem is that by the time the results were made available and discussed with the brewer, that particular batch of wort or beer will have long since moved on to the next stage. Any remedial procedures would only help subsequent brews.

Farsighted Brewers now use a quality assurance approach to plant hygiene, allied to the use of rapid microbiological methodology. Much more attention is given to plant design for easy cleanability, checking of the efficiency of cleaning (CIP) systems (e.g., caustic checks), confirming that the pasteurizer is working by testing temperature and applying various checks to test that heat-sensitive components are being destroyed, and so on.

Various rapid microbiological techniques have been advocated. The most publicized and most widely used is based on "ATP bioluminescence." The method depends on the firefly, an insect that emits light from its tail as a mating signal. This reaction depends on an enzyme called luciferase, which converts the chemical energy store found in all organisms (ATP) to light. The enzyme can be extracted and this reaction carried out in a test tube. The more ATP is present, the more light is produced, and it can be measured using a luminometer.

The rapid test used by brewers requires that a swab be scraped across the surface that needs to be tested. The end of the swab is then broken off into a tube that contains an extractant, together with the luciferase, and after a period that can be as short as a few minutes, the amount of light emitted is measured. The dirtier the surface (i.e., the more bugs and debris on it), the more ATP will have gotten onto the swab and, in turn, the more light will be measured. And so, in real time, an indication can be obtained of the state of hygiene of the plant. The method has been extended to measuring very low levels of microorganisms in beer, enabling the brewer to release beer to trade with confidence just a few hours (or, at most, days) after it was packaged.

SENSORY (ORGANOLEPTIC) ANALYSIS

Although the drinking of beer is a complex sensorial experience, bringing into play diverse visual stimuli and environmental factors (see chapter 4), ultimately it is the smell and taste of a product that will decide whether or not it will prove acceptable to the consumer. For this reason, much time and effort is devoted within the brewery to the tasting of beer at all stages in its production.

One of my past jobs was as the quality assurance manager of a large English brewery, where the first job at break of day was to stand alongside Neil Talbot, the head brewer, and taste the previous 24 hours' production. (This is not quite so delightful as it might sound, believe me!) A sip of each beer would be taken and "scored" on a scale of 1 to 4. A value of 1 indicated that the beer was of the expected high quality; 2 meant a minor flaw that would warrant a quick check of the records, but the beer could go to trade, as any deficiency was predicted to be imperceptible "in the trade"; 3 indicated a serious shortcoming in the product that demanded serious investigation and a holding of the beer while a decision was made about what to do with it; while 4 meant there was a major problem, the beer would have to be destroyed, and an urgent inquiry would have to be launched. Happily I don't recall any scores of 4 and very, very few 3's. We tasted beer at the cold conditioning stage, at the post-filtration stage, and after packaging. We also checked the water that was to be used to brew beer and dilute high-gravity beer.

The system was straightforward and highly effective as a screen to ensure that the highest quality standards were being maintained and that we identified as early as possible in the process if things were going awry. For instance, by tasting beer prior to packaging we could "nip in the bud" any faults before the expensive packaging process was carried out. It demanded, of course, that Neil and I be sensitive to the flavors expected in each product. As a QA technique, it served the purpose for which it was intended.

Taste, though, is a complex sensation that depends on the interaction of beer components with many receptors in the mouth and on the no less complex aroma perception through the nasal system. Sweet, sour, salt, and bitter are the basic tastes contributed by any foodstuff, and there are receptors for each on the tongue. There are, however, many other flavors in a product such as beer; they are all detected ultimately by receptors within the nose, although it is now accepted that there are receptors for many of them on the tongue as well.

Because of this complexity, it is not surprising that drinkers differ considerably in their sensitivity to different flavors. People can be "blind" to certain characteristics or acutely sensitive to others. In either instance, it can be a problem. It's just as well that Neil and I seemed to be fairly "middle ground." If we had been incapable of spotting diacetyl, then we could have released to trade beer that most people would have deemed undrinkable. Equally, if either of us had been acutely sensitive to a given character (and I must admit to being just that with the butterscotch note from diacetyl, see chapter 9), then we might have rejected beers that the vast majority of the population would have judged perfectly acceptable.

For this type of reason, beer tasting can be much more sophisticated than simply having a head brewer and QA manager standing around a spittoon (not that we ever spit it out, because (a) it was too good, (b) there is no snob value in doing so, (c) what a waste). In truth, there is justification for swallowing. Some folks would have it that the bitterness receptors are particularly prevalent toward the back of the tongue, but, even more importantly, carbon dioxide comes out of the beer in the throat and drives aroma components into the nasal passages.

It is essential that reliable and statistically well-founded tests be available to provide authoritative and semiquantitative information that can be applied to make decisions about beer quality. Broadly, these methods can be divided into difference tests and descriptive tests.

DIFFERENCE TESTS. As the name suggests, these are intended to tell whether a difference can be perceived between two beers. For instance, the Brewer may be interested in checking whether one batch of beer differs from the previous production run of the same beer, whether a process change has had an effect on the product, whether batches of the same brand of beer brewed in two different breweries are similar, and so on.

It is essential that the tasters not be distracted in this task. The environment has to be quiet, and they must not be influenced by the appearance of the product, so the beer is served in dark glasses and in a room fitted with artificial red light and with no opportunity for them to make contact with other assessors. It is important that the sensitivity of the tasters not be influenced by their having recently enjoyed a cigarette or a coffee or partaken of any strongly flavored food; best to have the tasting session prior to lunch, especially if curry is on the menu.

The classic difference procedure is the three-glass test: a minimum of seven assessors is presented with three glasses. Two of the glasses contain one beer, the third the other beer. The order of presentation is randomized. All the taster has to do is indicate which beer she thinks is different. Statistical analysis will reveal whether a significant number of tasters are able to discern a difference between the beers, and therefore whether, according to the law of averages, two beers will or will not be perceived as tasting different by the public.

DESCRIPTIVE TESTS. The three-glass test can be carried out essentially by anyone. However, if a Brewer wants to have specific descriptive information about a beer, he must use trained tasters, people who are painstakingly taught to recognize a wide diversity of flavors, to be articulate about them, and to be able to "profile" a beer.

The terminology that is used is usually of the type illustrated in figure 4.1. A group of individuals will collect around a table and taste a selection of beers, scoring the individual attributes, perhaps on a scale from 0 (character not detectable) through to 10 (character intense). Obviously it takes real ability to be able to separate out the various terms and recognize them individually, without one parameter influencing another. Once the scoring is complete, the individuals will discuss what they have found and agree on how the flavor of a beer should be summarized.

This type of test is widely used to support new product development and brand improvement and, of course, to characterize beers from a competitor. Once again, there are variants of it, such as the trueness-to-type test. This latter procedure is well suited to assessing whether a beer brewed in one brewery is or is not similar to the reference (standard) beer brewed in another location. For each of various terms found in the flavor profile form, each assessor is asked to mark whether the sample has got the same degree of that character (score = 0), slightly more (+1), substantially more (+2), slightly less (–1), or substantially less (–2). Obviously, the more flavor notes that are judged to have a score of 0, the more similar are the two beers.

Another test applied by Brewers is the evening "drinkability" or "session" test. This is designed to assess whether a beer will prove satiating or whether the consumer will want to drink more than a single glass of it. One variant involves presenting the tasters (who don't necessarily have to be trained) with two or three different beers whose drinkability is under assessment. The drinkers are asked to sip each of them, pass comment on them, and then select one for continued drinking. They are able to switch to another beer at will.

A careful record is made of how much of each beer is consumed—the highest volume indicates greatest drinkability—and the drinkers sent home by cab.

Such a test is, of course, somewhat primitive and unsophisticated, even if it can be rather informative and, let's be honest, fun. Many Brewers would value a straightforward test that will tell them in an uncomplicated way: will the drinker *like* this beer? Alas, such a test must lie a good way into the future.

Which brings me now to ask: what does the future have in store for beer and brewing?

TO THE FUTURE

MALTING AND BREWING IN YEARS TO COME

The fundamental shape of the malting and brewing processes has remained similar for many years. The reader should not conclude from this that the industries are stagnant or primitive but rather should appreciate that the basic route from barley to beer, aided by hops and yeast, is essentially well fitted to the purposes for which it is intended. I hope, too, that a study of the previous chapters will lead the reader to conclude that an enormous amount of research has been devoted to unraveling the science of malting and brewing and to the application of this knowledge in making these processes living and thriving testimony to a time-honored biotechnology.

This author does not foresee a dramatic change in the unit processes of malting and brewing in the foreseeable future. Fundamentally this is for two reasons. First, the nature of beer is as it is *because* of these processes: its flavor, its foam, its texture, its color, its wholesomeness are all dependent on the care and devotion invested by the Maltster, the Brewer, and the suppliers of hops and other ingredients. Which leads me to the second justification for leaving the basic procedures as they are: Brewers *care*, they take a pride in their products and in their heritage and they are fundamentally convinced that the best interests of the consumer are served by ensuring that they adhere to professional standards. *Of course* the Maltster and the Brewer expect to operate efficient processes, using raw material and plant capacity resources economically. They know only too well, however, that their beers have the character they do because of a vast myriad of chemical and biochemical changes occurring during malting, brewing, fermentation, and downstream processing. It is a high-risk strategy to mess about with them. Accordingly, the farsighted Maltster or Brewer will listen attentively to suggestions for process adjustment and will apply the science conscientiously, but will resist absolutely any development that jeopardizes their product.

This book is filled with examples of how the malting and brewing processes have developed, and have become vastly more efficient, without fundamentally modifying the basic route from barley to beer. In chapter 5 we saw that interrupted steeping enabled the malting process to be foreshortened by several days and how the addition of extra gibberellic acid, a molecule naturally found in barley, can further speed up the process (if it is used, which it isn't in the United States). Chapter 7 tells that the essential bitter and aroma ingredients of hops can be introduced more efficiently into the process in a form free from the vegetative parts of the plant. Chapters 8 and 9 show how the brewhouse and fermentation operations have been subtly altered to enormously improve efficiencies, but without inherently changing the character of wort or beer; developments have included high-gravity brewing, pure yeast technology, diacetyl control, and enhanced yeast-handling strategies. Chapter 10 indicates how enormous attention has been paid to stabilizing beer, with beneficial effects on the consistency of beer quality; advances here have included sterile filtration, use of nitrogen gas, and the application of stabilizing agents such as PVPP and silica hydrogels to allow more rapid turnaround times, if that is the Brewer's desire. Finally, in chapter 11, we found how developments in analytical techniques are being applied by Brewers to achieve tight control over their process and product, with genuine benefits for the consumer.

The future will see more of these improvements in the processes occurring. Perhaps the most publicized opportunity centers on the use of gene technology.

GENE TECHNOLOGY

As I write, no Maltster or Brewer is deliberately and directly using genetically modified raw materials, with the exception of a Brewer in Sweden that is overtly using such ingredients and incurring the wrath of Greenpeace in the bargain. The question is: will Maltsters and Brewers take advantage of this exciting new technology? I believe the answer is *yes,* but *only* once they are absolutely convinced that there are real merits in so doing.

We have seen clear evidence of the readiness of these industries to embrace new technology, but there is also absolute caution applied by Maltsters and Brewers whenever change is suggested: only when justification is 100% will a move be made. One has only to survey the history of brewing science to realize the truth of this statement. It is now over twenty-five years since the first research on genetic modification of brewing yeasts took place, and plenty of yeasts have been successfully modified. And, as yet, *none* of them is in commercial use.

Only one genetically modified brewing yeast has been cleared through all the necessary authorities, this in the United Kingdom. Should a brewing company wish to use it, they may. As yet, none has taken up the option. In part this seems to be because no Brewer wishes to be first into the marketplace with a beer labeled "product of gene technology." More importantly, however, none seems to be convinced that the merits of this particular organism outweigh the very real concerns that exist with the application of this science. The first Brewer to employ genetically modified yeast will do so because it brings genuine benefit to the consumer. Perhaps the yeast will boost the levels of some component of beer that is beneficial to health (see chapter 4). Or might it be a yeast that enables beer to be brewed substantially more cheaply—although the author fails to see how the science of genetic modification can hope to address one of the biggest cost components of beer in many countries: excise duty! (See below.)

The one yeast so far cleared for commercial use was "constructed" in one of my former research teams, led by John Hammond and his colleagues. Into a lager strain was introduced a gene from another yeast, this gene coding for an enzyme that will convert more of the starch into fermentable sugar, thereby enabling more alcohol to be made per unit of malt or, alternatively, enable less malt to be used per unit of alcohol. As we saw in chapter 8, not all of the starch from barley is converted into fermentable sugars in conventional brewhouse operations. To produce the so-called diet beers that have more (even all) of these partial degradation products of starch (dextrins) shifted into alcohol, Brewers add an enzyme (called glucoamylase) that is capable of performing the extra conversion. What we did was to take the bit of the genetic code from *Saccharomyces diastaticus* that codes for this enzyme and transfer it to a conventional bottom-fermenting strain of *Saccharomyces pastorianus*. This was done so efficiently that the extra DNA stayed in the yeast from generation to generation. Most importantly, we had transferred DNA from an organism that was extremely similar to the host organism: from one yeast to another one. And it worked! The host yeast was able to make the enzyme from the "foreign" bit of DNA and spew it into the wort, and there it chopped up the dextrins. The fermentations were performed on scales as large as 100 hectoliters, and the beer produced was indistinguishable from that produced conventionally. The beer was produced, bottled, and labeled for research purposes only.

The genetically modified yeast employed in making Nutfield Lyte was used as a test case for the purposes of seeking approval from the necessary U.K. authorities and advisory groups. For approval to be granted, the yeast had to be cleared by *four* entities: the Advisory Committee on Novel Foods and Processes within the government's Ministry of Agriculture, Fisheries and Food; the Advisory Committee on Genetic Modification (part of the government's

Health and Safety Executive); the Advisory Committee on Releases to the Environment (Department of the Environment); and the Food Advisory Committee. Four different departments had a say—four separate elements to scrutinize every facet of the science, ethics, and safety of the project and who had to be satisfied before permission was granted. And *still* this yeast remains in the freezer awaiting application. Everybody is applying understandable caution, but all the evidence is that the technology is sound and safe, provided that a responsible attitude is adopted.

This is certainly the case for Brewers, and for Maltsters, too, although there is some distance to go yet before suitable genetically modified barleys become available. They *will* be developed—with "new" properties such as enhanced disease resistance, enabling a reduced need for spraying with pesticides. The Maltster will adopt the same cautious approach as the Brewer on whether to use them. Of course, both the Maltster *and* the Brewer have a stake in the use of barley; indeed, ultimately it will be the Brewer that will drive the use or otherwise of genetically modified barley.

Gene technology, then, is an exciting concept and one that could provide genuine benefits. All the signs from the brewing industry are that the technology will only be used if those benefits accrue to the consumer.

WHAT WILL THE INDUSTRY LOOK LIKE IN TEN YEARS?

So how will our beer be made in the future? Can we anticipate a radically different approach to the traditional and semitraditional processes that have been used to make the world's favorite beverage for thousands of years? Or will the basic shape of the business stay as it is, with incremental improvements rather than radical alternatives continuing as the status quo?

Some while ago I canvassed a selection of other international experts from within the malting and brewing industries, asking them how they saw matters unraveling over a ten-year time frame. Their (and my own) views can be distilled as follows.

RAW MATERIALS

Nobody envisages a dramatic shift in the grist materials that will be used for brewing. Indeed, a number of Brewers have shifted back from sizable use of adjuncts to grists that are largely (if not entirely) of premium malted barley. They are convinced that this offers genuine quality, though there remains a clear justification for using other cereals where they offer unique attributes to a product, in terms of flavor or color, for instance.

The contribution of malt and adjuncts to the cost of beer is relatively low. There really is little strategic or financial justification for taking shortcuts with them, unless they are not available (e.g., the banning of malt imports in Nigeria) or there is a further financial incentive so to do. Just such a case arose in Japan, with taxation legislation that led to disproportionately less duty on "sparkling malt drinks," otherwise known as *happoshu*. They must contain less than 25% malt. They are packaged like beers and, to the consumer, are clearly from the same stable, even if the label cannot use the word "beer." The reality is that savings on a tax bill will be much more significant than savings on the grist bill. The reader will note that the Japanese brewers have not strayed from their traditional high-malt recipes for their flagship products.

The belief is that pressures will continue to minimize the use of additives in the growing of barley and its subsequent malting, yet everyone realizes that these agents can offer real advantages to the process and product; better to use a pesticide in the production of barley and to ensure its removal during steeping than to run the risk of a fungal infection of grain. Here, too, may be a major target for genetic modification: the construction of barleys that have inbuilt resistance to attack by undesirables.

There is concern that not enough premium malting-quality barley will be available to meet the increasing demand for it. And this situation is exacerbated considerably by the shift to growing corn as a source material for the production of grain alcohol destined for automobiles. The surge of bioethanol has thrown the grains market into turmoil—suddenly the growing of a prestigious but high-maintenance crop like malting barley seems unattractive to many. Leading hop varieties, particularly those with good aroma characteristics, will continue to be in heavy demand, and there will be shortages. The shortage may be exacerbated if hop growers succeed in their quest for alternative uses for hops, for example the identification of a high-value phytonutrient or two, which would suddenly rocket the value of hops. The cost of hops has already shown dramatic inflation as I write (March 2008). As brewers got ever more adept at extracting the last drop of bitterness from the crop, at a time when the world trend (despite the efforts of the U.S. craft brewers with their mega-bitterness beers) is toward *less* bitterness, so too did hop growers find themselves with surplus capacity. They responded by grubbing out hop yards and growing something more promising. The tipping point was reached in 2007, when there was insufficient hop supply to satisfy the global beer demand. Bigger brewers with their forward purchasing contracting did not suffer unduly, but many a smaller brewer reached for the telephone and called around in desperation, trying to locate hops.

BREWING

The crystal ball suggests that brewhouse operations ten years hence will not be radically different from those in place today. Already there has been an increase in the number of breweries incorporating mash filters rather than lauter tuns (see chapter 8), and no successor to the mash filter seems to be emerging. Perhaps brewhouse operations will become continuous, to match continuous fermentation operations, and yet few if any brewers seem to be convinced that continuous production of beer would be right for them.

Without doubt, though, Brewers, just like any other industrial sector, seek to lower their cost base. They all share the opinion, for instance, that processes will become far more automated, taking advantage of the rapid developments that are being made in the miniaturization, sensitivity, and flexibility of information technology. Automation has already happened to a certain degree, with substantial reduction in workforces having occurred over the years. The most impressive example of automation I have seen was in the warehouse operations of a major Japanese brewery. All but one of the forklift trucks was a robot, each busily shifting beer around the site according to a pre-program. And each truck played a tune as it trundled along—one was whistling *Yankee Doodle!*

PACKAGING

One brewer suggested to me that brewing could evolve to be a service to a distributed packaging industry, in just the same way that the soft drink industry operates today. It is certainly the case that the bulk of the raw materials cost, approximately half of the cost of processing, and, indeed, much of the innovation, is at this stage in the brewery operation. Therefore, issues such as reduced raw materials costs (for instance, use of aluminum), recycling, de-manning of what tends to be a very labor-intensive function, and energy conservation are all to the fore, as, of course, is the trend toward plastic instead of glass bottles.

THE PRODUCT

The common theme, however, which ran right the way through the replies on raw materials, brewing, and packaging and into consideration of the beer itself, is *quality*. In particular, Brewers anticipate beers having extended shelf lives to meet longer distribution chains, and there will be much more choice. The consumer is becoming more enlightened about issues of wholesomeness and quality; Brewers appreciate that they will have to meet drinker demand in this area, including the development of new beers with unique properties based around variables such as flavor, foaming, texture, and, in a very responsible manner, health attributes.

Research into so-called Consumer Sciences is developing fast. In the future, it will be possible for producers of all types of foodstuffs, including beer, to be able to forecast with much more confidence which products will be enjoyed by the customer. Understanding of the specific effects different components of beer have on the sensory apparatus in the mouth and nose will enable beers to be "designed" that are best suited to the enjoyment of the consumer.

INPUTS AND OUTPUTS

Brewers will continue to strive toward minimizing inputs; for example, they will continue to develop their processes toward less energy demand and less water. Consumption figures for efficient and inefficient breweries, respectively, are shown in table 12.1. Many breweries have headway to make to catch up with the pack. Even the leaders in the efficiency stakes are eager to improve.

The really energy-demanding operations in malting and brewing are malt kilning and wort boiling, and there is still some way to go in making substantial savings here without jeopardizing product quality. Developments focus on heat recovery, of course. The cost of refrigeration is also a major factor, hence the interest of a number of Brewers in lessening chilling demands.

There is some concern regarding the ongoing availability of good quality water for malting and brewing, forcing yet more attention on reducing water consumption in cleaning operations and on the need for more recycling. Concerning outputs (other than the beer itself, of course), Brewers again are adjusting their processing to reduce wastes such as spent grains, surplus yeast, and, of course, wastewater. Naturally, the more efficiently a material such as malt is converted into beer, the less overflow there will be to spent grains. There is a long, long way to go, though, before malt will be entirely convertible into wort and thence beer. The furthest anyone has got within the remit

TABLE 12.1. INPUTS FOR THE PRODUCTION OF 100 HECTOLITERS OF BEER

Input	Efficient brewery	Less efficient brewery
Malt	1.5 tons	1.8 tons
Water	500 hectoliters	2,000 hectoliters
Energy	15 gigajoules	35 gigajoules
Electricity	1,000 kilowatt-hours	2,000 kilowatt-hours

of a conventional brewing operation was my own previous organization (BRi), on a pilot scale, with the acid hydrolysis of the surplus grains to produce more sugars. The problem became one of "spent spent grains": the husk of barley is pretty resilient! Hull-less varieties of barley are available, though they are susceptible to infection in the field and are difficult to malt because the naked grains tend to stick to one another. In terms of saving water, the focus has to be on the amount required for washing and cleaning purposes. Naturally, the smaller the plant relative to the output of beer, the less cleaning water is needed. Indeed, continuous processes run for days or weeks, potentially years, without stripping down and cleaning, and are therefore very economical in terms of water usage.

A former colleague was fond of drawing attention to the apparent illogicality of the malting and brewing processes: "We take moist barley from the field and, in drying, heat it to *drive off* water. Then we *add* water in steeping, germinate, and then *drive off* water in kilning. To the brewery—and we *add* water in mashing. Then we *drive off* water in boiling..." I took the point, of course, but reminded him that all of these stages are performed for very good reasons, which is not to say that there may not be radical alternatives in the future.

Some folk point out that a goodly proportion of the dry weight of a barley kernel never finds its way into a beer. What a waste, they say. They forget that good, efficient modification of the endosperm requires embryo growth—it's the price to be paid. OK, then, comes the reply, use raw barley and tip in the enzymes from a bucket. Possible, indeed it has been done—but the flavor is *not* as good and, anyway, the extra cost of processing in the brewery (e.g., shorter filter runs) takes away much if not all of the financial benefit. Not to be deterred, the revolutionist shifts to an argument for taking the cheapest alcohol source one can find on the spot market and tipping in the flavor, color, and foam from a bottle. Sure, you can do it—but is it beer? And what will you do with all the surplus mash tuns, lauters, kettles, whirlpools, fermenters, and sundry other items? I have no difficulty with the research being put in place to do this—but for me it's a technology that will only really have its zenith on board interstellar craft headed on centuries-long journeys into outer space—and long after we traditionalists have departed this mortal coil! Beer is increasingly marketed on a platform of care, tradition, and benefit. That ought to mean that we don't stray too far from the present way of doing things (which has only truly been tweaked in relative terms over many generations) unless other substantive pressures come to bear. The fact that Brewers spend over $2.50 per barrel on advertising as opposed to just a few cents on scientific research ought to give the reader a reasonable grasp of what is generally considered to sell beer.

THE INDUSTRY

There will be an ongoing drive toward international brands, recognized names thriving far beyond home base. This will be achieved by acquisition, joint ventures (of the type seen in the construction and modernization of breweries in China), and brand licensing and contract brewing. Brewing companies will become further polarized, into the ever bigger at one extreme and the very small at the other; it will be those in the middle order that will find survival ever more challenging. More and more beer will be consumed at home, which is one of the justifications for increasing the shelf life of the product: once a beer has been retailed, the Brewer has no further control over its handling. All he can hope to do is build robustness into the product.

"Robustness into the product": those are apt words, indeed, with which to bring this book to a conclusion. Brewers (and their colleagues, notably Maltsters and hop suppliers) have devoted themselves for many, many years to delivering to the public a wholesome and flavorsome product, robust and so very consistent, glass to glass.

Beer has a long and proud tradition. Thanks to more than a century of dedicated research, brewing has developed into a tightly controlled, efficient technological process, albeit one, unavoidably and fascinatingly, subject to the vagaries of its agricultural inputs.

Brewing is very much a science. Engage a brewer in conversation, though, and see the twinkle in his or her eye and you will rapidly come to the conclusion that they love their brewing and their beer—just as any connoisseur loves their chosen art.

APPENDIX

SOME SCIENTIFIC PRINCIPLES

I realize that not everybody reading this book will have enjoyed a scientific training—and some of those who did will not necessarily have enjoyed it and, as a consequence, will have blotted it out of their memory banks. To help such people I offer here a simple crash course in chemistry, biology, and biochemistry, with just a little physics thrown in for good measure.

ELEMENTS AND COMPOUNDS

All matter in this world, whether animal, vegetable, or mineral, consists of chemicals. One of the simplest of these—and yet one of the most important—is water. Most of us know it by the *formula* H_2O, which means that it is made up of two hydrogen *atoms* and one oxygen atom. An atom is the smallest unit of an *element*, and it is from the elements that all matter is composed. At the last count there were over 100 elements. The simplest is hydrogen, which is given the symbol H. Other important ones include oxygen (O), nitrogen (N), sulfur (S), sodium (Na—after the Latin *natrium*), and chlorine (Cl). Perhaps the key element in life is carbon (C). So-called *organic chemistry* is the chemistry of carbon *compounds*. The key components of living organisms are *organic compounds*, which means that carbon is a key element in them.

A compound is a chemical entity, with its own individual properties, that comprises a collection of atoms of the same or different elements. The basic unit of a compound is a *molecule*. Water, H_2O, is a compound of hydrogen and oxygen. So, too, is hydrogen peroxide, H_2O_2. The latter molecule has just one extra oxygen atom, but this makes all the difference. Hydrogen peroxide is extremely reactive and finds uses as diverse as bleaching hair and sending rockets to the moon, not to mention stimulating barley to germinate. Water is, of course, a wonderful *solvent* (a solvent is something in which a *solute* can *dissolve*: for instance if you add sugar to water, sugar is the solute and water the solvent). As we shall see, it is this ability to dissolve things that makes water so important to life—including the brewing of beer!

There are a great many organic compounds. Some are very simple—for example, natural gas comprises the simplest, *methane,* whose molecule consists of one carbon and four hydrogen atoms (CH_4). (Incidentally, carbon dioxide, CO_2, the compound that puts the fizz in beer, is *not* classified as an organic compound.) At the other extreme are very complicated molecules that consist of a great many carbon atoms, and other atoms, too. Here I will refer to those compounds that are relevant to living systems, such as barley, hops, and yeast.

CARBOHYDRATES

Starch is a carbohydrate; so too is sugar. In fact, the term *sugar* refers to a wide range of related substances and not just *sucrose,* which is the granulated sugar that you stir into your coffee. The simplest sugar is *glucose,* which has the formula $C_6H_{12}O_6$. Like other sugars, it is very sweet. Glucose can exist in an open chain form and as two closed rings, in which the –OH group at carbon atom number one (C_1) is pointed downward (α) or pointed upward (β) (figure A1). I show two ways of drawing glucose, in the second of which all but one of the carbon atoms have not been shown. This is standard practice when organic chemists draw formulas. Every time you see two lines join or a line end without another type of atom (e.g., H) signified, there is a carbon atom at that point.

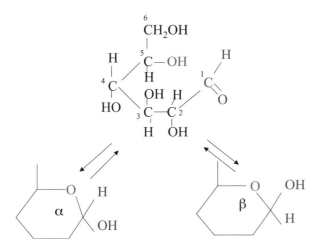

FIGURE A1 Glucose can exist in an open chain ("linear") form or as a closed ring, formed by an interaction between the groupings at carbon atoms numbers 1 and 5. The ring can be formed in two ways, with the –OH at C_1 pointing downward (alpha) or upward (beta).

Sugars such as glucose are able to join together to make bigger molecules. They do this by splitting out a molecule of water between them. If two glucoses join together, they make *maltose* if the –OH from the C_1 atom is lowermost or *cellobiose* if the –OH from the C_1 is uppermost (figure A2). Now, if water is added to maltose or cellobiose, then the *reverse* reaction can occur, and it will be split into two glucose molecules. When water is used to break up molecules in this way, the reaction is called *hydrolysis*. Usually this doesn't happen spontaneously; for instance, if you dissolve maltose in water, very little of it is hydrolyzed to glucose. Maltose needs help to be broken down, and this help comes in the form of enzymes, which we will come to shortly.

Maltose can pick up another glucose, and the resultant sugar is malto*triose* (*tri* indicating that this molecule contains *three* glucoses). Add a fourth glucose and you have maltotetraose, and so on. Each of these molecules, with one, two, three, four, or more glucose units has different chemical properties. For example, they are progressively less sweet.

Molecules containing relatively short chains of glucose units are known as *dextrins*. Sometimes they are called *oligomers,* and the basic building block, glucose, is called a *monomer* (mono- = "alone"). When there are lots and lots of building blocks, in this case glucose, linked together, we have a *polymer.* In the case of polymerized glucose, the best-known molecule is *starch,* which is the major food reserve in the barley grain. Polymers of sugars are called *polysaccharides*; the building blocks (in this case glucose) are called *monosaccharides*; dextrins are *oligosaccharides*.

Starch is an α-glucan. If, however, the links have a β-conformation, then the resultant polymer is a β-glucan, which has a totally different set of properties from starch. The latter is the main component of the barley cell wall, for instance, whereas starch is the food reserve packed within those cell walls.

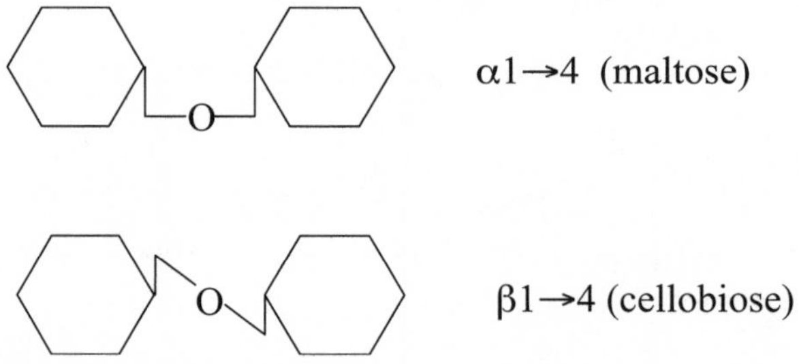

α1→4 (maltose)

β1→4 (cellobiose)

FIGURE A2 If a glucose in an alpha conformation reacts with another glucose with the splitting out of water between carbon atoms 1 on the first residue and 4 on the second, an α1→4 bond is produced. If the glucose is in the β-conformation, then an entirely different molecule is produced, with a β1→4 linkage.

PROTEINS

A different type of polymer found in living systems is *protein*. Here the monomer is not sugar but *amino acid*. These are simple molecules that, unlike sugars, contain nitrogen atoms. There are twenty or so different amino acids, each of which has different properties. Just like sugars, they can link together by splitting out water to form long chains. Again, as for polysaccharides, the reverse process can take place, and addition of water to a polymer of amino acids leads to hydrolysis to individual amino acids.

When a few (say, 2–10, although the exact definition is somewhat arbitrary) amino acids are linked together the molecule form is called a *peptide*. A molecule containing, say, 10–100 amino acid monomers is called a *polypeptide*. Bigger molecules are called *proteins*.

Because there are different amino acids, they can be linked together in many different sequences, and thus there is tremendous diversity between proteins in their structure and properties. Egg white is composed of protein; so too are the nails on your fingers and the silk in your tie or your dress. They are all very different.

The proteins in nails or silk are *structural* proteins. Another very important class of proteins is the *enzymes*. These are found throughout biological systems and are *catalysts*. A catalyst is a compound that speeds up a chemical reaction or enables it to take place. For example, we saw above that if two glucoses link together they form maltose. This joining together doesn't happen spontaneously; it has to be catalyzed. There is an enzyme that does that job. There is also an enzyme, a different enzyme, that enables the reverse reaction, the breakdown of maltose into glucose by the addition of water. Because it catalyzes a hydrolysis it is called a *hydrolase*. Wherever you see the suffix *-ase*, it refers to an enzyme. Synthases catalyze synthetic reactions, such as the coupling of two molecules together. Decarboxylases split carbon dioxide out of molecules. Oxidases add oxygen to molecules, whereas dehydrogenases take hydrogen out of molecules. There are many different enzymes, each of which has its specific job to do.

For the most part, an organism will only produce an enzyme when it needs it to do a job of work. Take barley, for instance. When it is ready to germinate, it needs to break down the food reserves in its starchy endosperm. First of all, it needs to break down the cell walls that are the wrapping in the endosperm, so it needs first to make the enzymes that do this job. Key among these enzymes are the *β-glucanases*. So these are the first enzymes to be produced by the protein synthesizing machinery in the aleurone tissue, which responds to a specific hormone trigger from the embryo (see chapter 5). (*Hormones* are molecules, usually small ones, which signal that a specific change needs to take place in a living system. They don't effect that change themselves.)

Once the walls are gone, then the barley needs the *proteinases* to break down the proteins, exposing the starch, which in turn will be hydrolyzed by the *amylases* (as we saw in chapter 5, the key components of starch are amylose and amylopectin, hence the name of the enzymes that degrade them).

Amylases, β-glucanases, and proteinases are different enzymes. They each have their own job to do. By ensuring such specificity in enzymes, living organisms can maintain control over their *metabolism* (an organism's metabolism is the sum total of all the reactions involved in its life cycle).

To act, enzymes need to get close to the molecules that they act on (the *substrates*). The reaction will generally occur only when there is water present within which the enzyme and substrate can move around. This is the reason why water must be introduced into barley before its metabolism can swing into action, but then driven off from malt in kilning when the maltster wishes to stop the modification process. It's also why you must add water to milled malt to get the starch hydrolysis reactions going.

All chemical reactions take place more rapidly at higher temperatures, the rule of thumb being that the reaction rate doubles for each 10°C rise in temperature. This applies to enzyme reactions, too, but there is a complication. Enzymes are, to a greater or lesser extent, inactivated by heat. Some are very sensitive and are rapidly killed at relatively low temperatures such as 50°C (122°F). The β-glucanases are an example of this high *lability*. Other enzymes are more robust—for example, some of the peroxidases (enzymes that use hydrogen peroxide as a substrate) in barley happily survive 70°C (158°F). These varying sensitivities to heat have major implications for the malting and brewing processes, as we see, for example, in chapters 4 and 8.

OTHER COMPLEX MATERIALS IN LIVING SYSTEMS

There are two other key classes of complex molecules found in all living systems, including barley, hops, and yeast: the nucleic acids and the lipids. It is not necessary to go into any great detail here, but suffice to say that the *nucleic acids* are the molecules involved in synthesizing the proteins of an organism. They contain the genetic code that determines the nature of an organism— whether it is a man or an amoeba, a barley or a hop, a yeast or an organism intent on spoiling beer. There are two types of nucleic acid: deoxyribonucleic acid (*DNA*) and ribonucleic acid (*RNA*). The former comprises the genetic code, or blueprint; the latter provides the protein synthesizing machinery.

There are many types of *lipids* in cells. Most of them contain very long chains of carbon atoms linked to hydrogen, so-called fatty acids. Their only property that I want to mention here is that, by definition, they are not soluble in water, but rather in other types of solvent. There is a saying in chemistry that *like dissolves like*. Lipids dissolve in so-called *organic solvents*. This is essentially the definition of a lipid. I refer you to the home for the simplest explanation. Think of a sugar such as glucose—you will see from its formula above that it has lots of –OH groups on it, rather like water, H_2O, or, if you like, H-O-H. Glucose readily dissolves in water. Now think of a greasy spot on your clothes caused by butter. Butter is composed of lipid, and you won't get rid of that stain by washing with water. You will need a solvent that also has a long carbon chain, something like petroleum. Because of this insolubility, lipids tend to be associated with structural elements in a living organism, and in a process such as brewing, they tend to associate with particles, such as the spent grains. Cooking fats, lipstick, and glass-washing detergents are all lipids, because of their water insolubility, and if they get into beer they will tend to go into the foam rather than the liquid beer. Once in the foam they disrupt it—and kill it (see chapter 4). My wife can pretty much deliver the funny lines from my classes, especially this one: if you're dripping fat off your mustache you will have lousy foam on your beer; if you are wearing lipstick, you will have just as big a nightmare; if you have a mustache *and* lipstick, then you have real problems.

CELLS

The fundamental unit of all living organisms is the cell. Some organisms, such as brewing yeast, are *unicellular*, in that they consist of just one cell. Organisms such as barley and hops are *multicellular*, with many different types of cells. Thus, in barley, there are embryo cells, aleurone cells, starchy endosperm cells, and so on. Collections of similar cells (e.g., the aleurone) are called *tissues*.

The bacteria that can contaminate wort and beer are also unicellular, but they are even simpler than yeast. In a bacterium, there is no division of the contents of the cell into compartments; all of the nucleic acids, carbohydrates, proteins, and other simpler molecules involved in the metabolism of the bacterium are in a watery soup called the *cytoplasm*. Such simple organisms are called *prokaryotic*.

The cells of yeast and other *higher* organisms (such as barley and hops) are *eukaryotic*: the cytoplasm is divided into distinct regions, called *organelles*. Just as the organs of the human body have their own roles, so too do the organelles within a cell. These are referred to for yeast in chapter 9.

Living cells need a source of energy, which, when released, is used by them to survive, to grow, to divide, or to do the job allocated to that cell. A cell in the embryo of barley will consume energy in making the hormones that it will send out to the aleurone cells, which in turn consume energy in producing the hydrolytic enzymes that will degrade the starchy endosperm. In organisms such as yeast and the barley grain, the energy is obtained by "burning" sugars: in a series of enzyme-catalyzed reactions, the sugar is degraded and energy is progressively released. The standard equation for *respiration* is given in chapter 9, which also indicates the equation for when the process is carried out in the absence of oxygen (*fermentation*). In both instances, the energy is collected in the form of a chemical carrier called *ATP*, which is found in all living cells and is often called the "universal energy currency." ATP is then used by the energy-consuming reactions, such as movement of cells, synthesis of new proteins and membranes, and the like.

FOOD RESERVES ARE POLYMERS—WHY?

Wouldn't it be easier if the cells of the starchy endosperm of barley were packed full of glucose and amino acids rather than starch and protein, meaning that all the embryo had to do was open up the wrapping cell wall and then bathe in the flood of goodies that would surge out? Yes it would, but it isn't possible, just as yeast must keep its food reserve, glycogen, in a polymeric form.

We have to understand the phenomenon of *osmosis* to appreciate the reason for this polymeric storage. If you have two liquids, one a concentrated solution of glucose and the other a dilute solution of glucose, and you separate them by a membrane, then water will progressively pass from the dilute solution to the more concentrated one until the strength of the solutions is identical on either side of the membrane. This is osmosis. The numbers of glucose molecules on either side of the membrane is critical in this experiment. Now, if those glucoses were all linked together as starch (see above), then instead of having many molecules of sugar in the concentrated solution, we would just have a single molecule. There is the same amount of sugar, but far less *osmotic pressure*. Herein we find the reason for the polymeric form of food reserves: if all the glucose and amino acids in the starch and protein food reserves of barley were monomeric, they would exert an enormous osmotic pressure in the cells, and water would flood into them and burst them.

pH

pH is a measure of the relative acidity or alkalinity of a solution. Although there are several definitions of an acid, for our purposes it's sufficient to say that it is a chemical substance that releases *hydrogen ions*. pH is a measure of the concentration of hydrogen ions; it might seem counterintuitive, but the lower the pH, the more hydrogen ions are present and the more acid is a solution. The symbol for the hydrogen ion is H^+. It has one positive *charge*. Ions are basically chemicals that have charges. They attract or repel other ions: one positive ion will repel another positive ion, but it will attract a negative ion. The saying goes, "Like charges repel, opposite charges attract."

One negatively charged ion is the hydroxide ion, OH⁻. If a hydrogen ion and a hydroxide ion get together by attraction, they can go as far as to react with one another and make...yes, water!

$$H^+ + OH^- \rightarrow H_2O$$

Clearly, if all of the hydrogen ions in a solution are mopped up by hydroxide ions, then the solution is neutral and not acidic. Its pH is 7.0. If there are more hydrogen ions than hydroxides, then the pH is below 7 and is acidic. The lower the pH, the more acidic is a solution. If there are more hydroxide ions than hydrogen ions, the solution is alkaline (caustic) and has a pH above 7.0.

Beer is acidic, with a pH usually between 3.8 and 4.6.

Hydroxide isn't the only negatively charged species that the hydrogen ion can react with. Others include the bitter substances, the iso-α-acids, which can exist in a charged, negative state at higher pHs, but when the pH is low (H⁺ is high) they "pick up" this ion and the charges cancel out, which means they become uncharged. This type of interaction is tremendously important. In the case of these hop compounds, it influences their bitterness and foaming properties. The uncharged forms are much bitterer and more foaming, and also better able to kill microorganisms, than are the charged forms. It is for this reason, too, that enzymes are more or less active at different pHs, because they too can have different structures depending on the extent to which their negative groups interact with the hydrogen ion.

Buffers are materials that can chemically "soak up" hydrogen ions and therefore stop a pH from changing. They are very important in living cells, because the cell's machinery is designed to operate to best effect at a particular pH, which needs to be kept fairly constant.

The starchy endosperm cells in barley and malt, therefore, have their preferred internal pH. A mash of malt will have a pH of around 5.5, due to an "internal" buffering system (including some of the soluble proteins, polypeptides, and peptides) that holds the pH at that value. Quite a lot of acid needs to be added to drop the pH from that value. The pH falls during fermentation because the yeast uses up the buffering system (peptides) and because the yeast releases acid.

One factor involved in lowering the pH in a mash is the level of calcium in the water (liquor). It does this by reacting with phosphate from the malt, releasing hydrogen ions:

$$3Ca^{2+} + 2HPO_4^{2-} \rightarrow Ca_3(PO_4)_2 + 2H^+$$

COLOR

The color of a liquid such as beer, of our clothes, or of the cover of this book is due to the extent to which our eyes detect different types of light.

Light can be thought of as a vast collection of different waves, each of which has a different size (wavelength), measured in nanometers (nm; 1 nm = one thousand-millionth of a meter). Visible light is a collection of light waves of anywhere between 400 nm and 800 nm. Blue light is at the shorter-wavelength end, red light at the longer-wavelength end.

If you have a light source equally strong at all of these wavelengths, the light you see is vivid white. Conversely, if there were no light whatsoever, you would see black. You would also see black if somebody put a filter between your eyes and that light source, a filter that screened out light at all the wavelengths. If, however, that filter sifted out only the longer-wavelength light, then you would detect the light as being blue, because it is the shorter wavelengths that are reaching the eye. If the filter trapped the shorter wavelengths, then you would see red light emerging.

This is the basis of our seeing different colors. Paints and pigments are the color they are because they absorb a series of wavelengths of sunlight *other* than those that are associated with the color that they reveal to you. A green paint has the shade it has due to a selection of wavelengths of light that it doesn't filter out and that therefore enter your eye.

Many individual chemical compounds absorb light of specific wavelengths, and this is the basis for measuring them. Take our friends the iso-α-acids again: they absorb what is known as ultraviolet light, which is light of very short wavelength, beyond the blue light at the lower-wavelength end of the spectrum, light that cannot be detected by the human eye. By measuring the amount of such light absorbed by a solution of the iso-α-acids, one can deduce how much of these materials are present, because the more of a chemical compound is present in a solution, the more light it will absorb. To measure these bitter compounds, they are extracted into a solvent and light with a wavelength of 275 nm is shone through the solvent. This is done in a spectrophotometer, which is a device that can split up light into individual wavelengths and measure how much of each wavelength is "taken out" by a solution. Spectrophotometry, using a wide range of wavelengths appropriate to the chemical to be measured, is extensively used in industry, including the brewing industry. At the longer wavelengths (beyond the red end of the spectrum) we come across the near-infrared (NIR) region. A number of chemical species absorb radiation in this part of the spectrum, including water and protein. This is taken advantage of in NIR spectroscopy to measure materials very rapidly, notably in the screening of barley entering into a malting after harvest. NIR can also be used for in-line measurement of alcohol in the brewery.

CHROMATOGRAPHY

Another analytical technique of enormous value in the brewery is chromatography. Fundamentally, this involves the separation of mixtures by passing a mobile phase past a stationary phase. Substances differ in their preference for the two phases and are either held on the stationary phase or tend to move along with the mobile phase. When the chromatography is complete, the individual substances are detected in some way, perhaps by measuring their absorption of specific wavelengths of light (see above) or staining them with a dye. There are various types of chromatography; in gas chromatography, the mobile phase is a gas mixture and the stationary phase some type of solid in a column. High performance liquid chromatography differs in that the mobile phase is a liquid at very high pressures.

GLOSSARY

ABRASION   Damaging the part of the barley corn furthest from the embryo in order to stimulate activity of the aleurone and "2-way" modification

ABSCISIC ACID   A plant hormone that counters the action of gibberellin

ACCELERATED FERMENTATION   Fermentations carried out under conditions where they proceed more rapidly, for example, by operating at a higher temperature

ACID WASHING   Treating yeast with acid in order to kill contaminating organisms without destroying the yeast itself

ACROSPIRE   The developing shoot in germinating barley

ADJUNCT   A source of fermentable extract other than malt for use in brewing

AEROBIC   In the presence of oxygen

AGING   The holding of beer in order for it to be converted to the desired state for retail to the consumer

AGITATOR   A device for mixing the contents of a vessel, for example a mash mixer

AIR REST   Period employed during the steeping of barley in which water is drained from the grain bed to allow the access of oxygen to the embryo

ALBUMIN   Soluble protein class in barley

ALCOHOL   Alcohols are a class of organic compounds containing the hydroxyl (–OH) group. The principal product of fermentation by yeast is ethyl alcohol (ethanol). Other, "higher," alcohols are also produced in much lower quantities by yeast, and they are implicated in the flavor of beer.

ALCOHOLIC STRENGTH (ABV)   The amount of alcohol in a beverage, frequently referred to as ABV (alcohol by volume), in which the ethanol content is quantified in terms of volume of ethanol per volume of beverage.

ALE   A type of beer generally characterized by an amber color and traditionally produced using a top-fermenting yeast. (In medieval England, ale referred to unhopped beer, but this no longer applies)

ALEURONE   A tissue 2–3 cells deep that surrounds the starchy endosperm of the barley corn and is responsible for making the hydrolytic enzymes (hydrolases) that degrade the barley food reserves

ALPHA ACIDS   Resins from the hop that are the precursors of the bitter compounds in beer

AMINO ACIDS   Small molecules (there are around 20 different ones) containing nitrogen, which are the building blocks of proteins

AMYLASES   Starch-degrading enzymes

AMYLOSE   A linear polymer of glucose, which is a key component of starch

AMYLOPECTIN   The second key component of starch, differing from amylose in that it has branches

ANAEROBIC   In the absence of oxygen

ANEUPLOID   Indicates that an organism contains more than two copies of its genetic blueprint. Haploid organisms contain one copy, diploid organisms two copies, and polyploid organisms many copies. There is no agreed point at which aneuploidy becomes polyploidy

ANTIFOAM   A material added to fermentations to suppress excessive production of foam

ANTIOXIDANT   A material either native to a raw material or else added that serves to protect against the damaging influence of oxygen

AROMA HOPS   Hop varieties said to give particularly prized aroma characteristics to a beer

ASTRINGENCY   A drying of the palate

ATP BIOLUMINESCENCE   A technique for detecting microorganisms and soil by measuring the amount of light produced by the action of the enzyme luciferase acting on ATP present in the sample

AUTOLYSIS   The breakdown of a cell by its own enzymes

AUXILIARY FININGS   Agents used alongside isinglass to facilitate the settling of insoluble materials from green beer

AWN   The beardlike projection on a barleycorn

BARLEY   *Hordeum vulgare;* a member of the grass family and the principal raw material for malting and brewing worldwide

BARLEY WINE   A very strong type of ale of long standing in England

BARREL   A volume measure of beer (United States = 31 U.S. gallons = 1.1734 hectoliters; United Kingdom = 36 U.K. gallons = 1.6365 hectoliters)

BEADING   The formation of bubbles of carbon dioxide in a glass of beer and their rise to the top of the drink

BEER STONE   A precipitation of calcium oxalate in beer dispense pipes

BETA-GLUCAN (β-GLUCAN)   A polymer of glucose that forms the bulk of the cell walls in the starchy endosperm and causes several serious problems to the brewer if not properly broken down in malting and mashing

BETA-GLUCANASE (β-GLUCANASE)   The type of enzyme that hydrolyses beta-glucan

BIOLOGICAL ACIDIFICATION   A practice common in Germany in which microorganisms (lactic acid bacteria) are encouraged to grow in the process in order to increase the acidity (lower pH)

BIOLOGICAL STABILITY The extent to which a beer is able to resist infection

BITTER Pale ale served on draft

BITTERNESS A flavor characteristic customarily associated with beer; also the term used to quantify the content of bitter compounds (iso-α-acids) in beer

BOCK A type of lager-style beer

BOILING The process of vigorously heating sweet wort at boiling temperatures

BOTTOM FERMENTATION Traditional fermentation mode for lagers where yeast collects at base of fermenter

BRACTEOLES The leafy parts of a hop cone

BREAKDOWN Deterioration of a beer

BREAK POINT The stage during kilning of malt when the temperature of the air leaving the malt becomes identical with that entering the malt because all of the free water that is not inside the malt has been driven off. Whenever there is unbound water present, it will consume energy (latent heat) in order that it can escape, essentially as steam. If this free water is taking up the heat, the air coming off the kiln remains relatively cool.

BREWHOUSE The part of the brewery in which grist materials are converted into wort

BREWSTER A female brewer

BRIGHT BEER Beer post-filtration

BRIGHT BEER TANK The vessel to which a beer is run after filtration and before packaging. Sometimes called a fine ale tank

BROMATE Has been employed (as potassium bromate) in order to suppress rootlet development during germination of barley

BURTONIZATION Adjustment of the salt content of brewing liquor to render it similar to that of the water at Burton-on-Trent in England

CALANDRIA A device either internally or externally linked to a kettle and used for heating wort

CALCOFLUOR A substance that binds specifically to β-glucans and reveals them via fluorescence

CARBONATION The amount of carbon dioxide in a beer; also, the act of increasing the level of carbon dioxide

CARBOXYPEPTIDASE An enzyme in barley that hydrolyzes proteins by chopping off one amino acid at a time from one end

CARDBOARD An undesirable flavor note that develops in packaged beer on storage

CARRAGEENAN An extract of seaweed used to aid solids removal in the wort-boiling stage

CASK The traditional vessel for holding unpasteurized English ale

CELL The basic unit of any living organism

CELLAR The part of a brewery containing the fermenters and the conditioning vessels. Also, the part of a retail outlet (e.g., bar) in which the beer containers are stored

CELL WALL The outside of a cell, whose role is to maintain the shape of that cell

CHARCOAL   A material capable of adsorbing flavors and colors from liquids that it contacts. Used, for example, to treat liquor coming into a brewery

CHILLING   The cooling of liquid streams in a brewery, for example hot wort going to the fermenter or green beer passing to conditioning and filtration

CHROMOSOME   The form in which the genetic material of a cell (DNA) is held in eukaryotic cells

CIP (CLEAN-IN-PLACE)   An integrated and automated system of cleaning with caustic and/or acid installed in modern breweries

CLING   The adhesion of foam to the walls of a beer glass (also known as lacing)

COALESCENCE   The tendency of bubbles in beer foam to merge together and form bigger bubbles

COLD BREAK   Insoluble material that drops out of wort on chilling

COLD WATER EXTRACT   A measure of modification of malt based on the small-scale extraction of milled malt in dilute ammonia

COLLOIDAL STABILITY   The tendency of a beer to throw a haze on storage

COLOR   The shade and hue of a beer

CONDITIONING   The maturation of beer in respect of its flavor and clarity

CONTINUOUS FERMENTATION   A process in which wort is converted to green beer in a few hours by passage through a vessel holding yeast

CONVERSION   The stage in mashing when the temperature is raised to enable gelatinization of starch and subsequent breakdown of the starch by amylases

COOKER   A vessel in the brewhouse in which adjuncts with very high gelatinization temperatures are cooked

COOLER   A device (often called a Paraflow) in which hot wort flows in close contact (but separated by a very thin metal sheet) with a cooling liquid in order to bring it down to the temperature at which fermentation will be carried out

COPPER   The vessel (often called the kettle) in which wort is boiled with hops

CORN   Maize. The word is also used to describe individual grains of barley

CRABTREE EFFECT   The control mechanism that dictates that yeast ferments sugar rather than metabolize it via respiration if the sugar concentration presented to the yeast is high

CROPPING   The collection of the yeast that proliferates during fermentation

CROWN CORK   The crimped tops used on beer bottles

CULMS   The rootlets of germinated barley that are collected after kilning and sold as animal feed

CURING   The higher-temperature phases of kilning when flavor and color are introduced into malt

CYLINDROCONICAL VESSELS (CCVS)   Tall fermentation vessels with a mostly cylindrical body but a conelike base into which the yeast collects after fermentation

DARCY'S LAW   An equation that explains the rate at which liquid flows, for example in a lauter tun or a beer filter

DECOCTION MASHING   Practice originating in mainland Europe in which a mash is progressively increased in temperature by taking a proportion of it out of the mix and boiling it prior to adding it back into the whole

DESCRIPTIVE TESTS   Beer tasting protocols in which trained tasters describe the taste and aroma of beer according to a series of defined terms

DEXTRINS   Partial breakdown products of starch that consist of several glucose units and are not fermentable by yeast

DIACETYL   A substance with an intense aroma of butterscotch that is produced by yeast during fermentation but is subsequently mopped up again by the yeast

DIATOMACEOUS EARTH   The skeletal remains of microscopic organisms used in powder filtration of beer (also known as kieselguhr)

DIFFERENCE TESTS   Blind tasting procedures in which tasters (including the untrained) are asked to differentiate samples of beer

DIMETHYL SULFIDE   Compound that imparts a significant flavor to many lager-style beers

DIRTY WORT   Wort containing a high level of trub solids and which is therefore turbid

DISPROPORTIONATION   The passage of gas in beer foam from small bubbles to larger bubbles, leading to a disappearance of the former and increase in size of the latter

DISSOLVED OXYGEN   The amount of oxygen dissolved in a wort prior to fermentation or, more commonly, the amount dissolved (and undesired) in beer.

DORMANCY   The control mechanism in barley that prevents the grain from germinating prematurely

DOWNY MILDEW   A disease of hops

DRAFT BEER   Either beer in cask or keg, or sometimes unpasteurized beer in small pack

DRINKABILITY   The property of beer that determines whether or not a customer judges it worthy of repurchase

DRY BEER   Beer genre in which beverage contains relatively low residual sugar

DRY HOPPING   Traditional procedure in the production of English cask ales in which a handful of hop cones are added to the cask prior to shipment from the brewery

DUTY   Excise tax on beer

DWARF HOPS   Hops that grow to a lower height than traditional varieties

EAR   The head of a barley plant that holds the grain

EMBRYO   The baby plant in the grain

ENDO ENZYMES   Hydrolytic enzymes that chop bonds in the inside of a polymeric substrate (examples are α-amylase and β-glucanase)

ENDOGENOUS ENZYMES   Enzymes in the malting and brewing process that are contributed by the raw materials (malt and yeast)

ENDOSPERM   The food reserve of the barley plant

ENZYME   A biological catalyst comprising protein

ESSENTIAL OILS   The aromatic component of hops

ESTER   A class of substances produced by yeast that afford distinctive, sweet aromas to beer

ETHANOL   The principal alcohol in beer, which is the major fermentation product of brewing yeast and which affords the intoxicating property to the beverage. Originally called ethyl alcohol

EVAPORATION   A measure of the water loss during wort boiling

EXCISE   Tax on alcoholic beverages levied by government agencies

EXO ENZYMES   Hydrolytic enzymes that chop bonds at the ends of substrate molecules, thereby yielding small products generally assimilable by organisms (examples are β-amylase and carboxypeptidase)

EXOGENOUS ENZYMES   Enzymes added to the brewing process from outside sources (i.e., not from malt or yeast)

FEED GRADE BARLEY   Barley that yields a relatively low level of extractable material after conventional malting and mashing

FERMENTABILITY   The extent to which a wort can be used successfully by yeast to produce ethanol

FERMENTATION   The process by which sugars are converted into ethanol by yeast

FILTRATION   The clarification of beer. Sometimes people refer to the recovery of wort from spent grains as "filtration," but strictly speaking this is "wort separation"

FINGERPRINTING   The differentiation of yeasts (or barleys) by analyzing the pattern of DNA fragments produced from them

FININGS   Materials used to clarify wort and especially beer by interacting with solid materials and causing them to sediment

FLASH PASTEURIZATION   Heating of flowing beer to a high temperature (e.g., 78°C) for less than a minute in order to inactivate microorganisms

FLAVOR PROFILE   An expert semi-quantitative evaluation of beer flavor made by trained tasters using defined taste and aroma descriptive terms

FLAVOR STABILITY   The extent to which a beer is able to resist flavor changes (usually undesirable) within it

FLOCCULATION   The tendency of yeast cells to associate

FOAM   The head (froth) on beer

FOAM STABILIZER   Either endogenous materials (e.g., proteins from malt) that stabilize foam or materials added to beer to protect foam (e.g., propylene glycol alginate)

FONT   The unit on the bar that labels the draft beer being served from that tap

FRANCHISE BREWING   The brewing of one company's beer under license by another company

FREE AMINO NITROGEN (FAN)   A measure of the level of amino acids in wort or beer

FUNGICIDE   An agent sprayed onto crops such as barley and hops to prevent the growth of fungi thereon and therefore ensure that those crops are healthy and high-yielding and don't introduce any harmful materials into the brewing process

FUSARIUM   An infection of barley that can cause a beer made from that beer to gush

GALLON   A standard unit of beer volume (U.S. barrel = 0.8327 U.K. barrel)

GELATINIZATION   A disorganization and loosening of the internal structure of starch granules by heating, rendering the starch more amenable to enzymatic hydrolysis

GENETIC MODIFICATION   A process of modifying the genome of an organism by introducing specific pieces of DNA from an exogenous source

GENOME   The information code of a cell, held within DNA, which determines the nature and behavior of that cell

GERMINATION   The process by which steeped barley is allowed to partially digest its endosperm and the embryonic tissues to partially grow

GIBBERELLINS   Plant hormones, produced within the embryo of barley, which migrate to the aleurone and trigger enzyme synthesis

GRAVITY   The strength of wort in terms of concentration of dissolved substances, as measured traditionally using a hydrometer

GREEN BEER   Freshly fermented beer prior to conditioning

GREEN MALT   Freshly germinated malt prior to kilning

GRIST   The raw materials (malt and other cereals) that will be milled in the brewhouse. More loosely applied also to those adjunct materials that don't require milling (e.g., syrups to be added to the kettle)

GUSHING   The uncontrolled surge of the contents of a beer from the package after opening

HAMMER MILL   A mill that grinds malted barley down to extremely fine particles suited to a mash filter for subsequent wort separation, but not a lauter tun

HAZE   Turbidity

HAZEMETER   An instrument for measuring the clarity of beer, operating on the principle that particles scatter light. The more light is scattered, the more particles are present. Some hazemeters measure the amount of light scattered at right angles (90°) to the light beam shone at the particles. Other meters ("forward scatter" meters) measure the light deflected at 13°. The former type is sensitive to extremely small particles, the latter to big particles

HEAT EXCHANGER   Device for rapidly cooling down liquid streams, for example boiled wort. The hot liquid flows countercurrent to a cold liquid on either side of thin walls. Heat passes from the hot to the cold liquid

HECTOLITER   Measure equivalent to 100 liters

HEMOCYTOMETER   Microscope-based chamber for counting yeast cells

HIGH-GRAVITY BREWING   Technique for maximizing vessel utilization whereby the wort being fermented is more concentrated than necessary to make the desired strength of beer. After fermentation, the beer is diluted to the required alcohol content.

HIGH PERFORMANCE LIQUID CHROMATOGRAPHY   Analysis technique involving separation and measurement of components of a mixture on the basis of their affinity for a high-pressure liquid stream or a solid support

HIGH-TEMPERATURE MASHING   Mashing performed at higher than normal temperatures in order to rapidly eliminate one of the starch-degrading enzymes (β-amylase) and produce a wort that contains fewer sugars that are fermentable by yeast and hence a lower-alcohol beer.

HIGHER ALCOHOLS Compounds similar to ethanol but containing more carbon atoms. They may contribute to the aroma of beer (and certainly do after conversion into their equivalent esters), and it has been suggested that they may be responsible for hangovers, although there is very little evidence for this

HOP Plant that provides bitterness and aroma to beer

HOP BACK Vessel rarely found these days that was used to separate boiled wort from residual solids by passage through a bed of waste hops

HOP CONE The flower of the female hop plant, which is the part of the plant used in the brewing process

HOP GARDEN Where hops are grown

HOP OIL The component of hops providing aroma (essential oils)

HOP POCKET A large sack packed with hops

HOP PREPARATIONS Extracts of hops, usually made with liquid carbon dioxide, that can be used at various stages in the brewing process to introduce bitterness or aroma to wort or beer more efficiently

HOP RESIN The precursors of bitterness in beer (α-acids)

HOPPED WORT Wort after the boiling stage

HORDEIN Insoluble storage protein in barley that is broken down during malting and mashing

HORMONE Small molecule that switches on or off events in a living organism; for example, gibberellins are hormones that switch on enzyme synthesis in barley

HOT BREAK Insoluble material that drops out of wort on boiling

HOT WATER EXTRACT A measure of how much material can be solubilized from malt or an adjunct, obtained by carrying out a small-scale mash of the material and measuring the specific gravity of the resultant wort

HUSK The protective layer around the barleycorn

HYDROGEL Material derived by acid treatment of silica that is used for the removal of potential haze-forming materials from beer ("chill proofing")

HYDROLYZED CORN SYRUP Material produced by the acid or enzymatic hydrolysis of corn starch and which can have different degrees of fermentability. Added to the wort kettle, thereby providing an opportunity to extend brewhouse capacity by avoiding the need for mash extraction and separation stages

HYDROMETER Device operating on a principle of buoyancy for measuring the specific gravity of a solution: the higher it floats the more material is dissolved in the solution

HYDROPHOBICITY A measure of the extent to which a molecule moves away from water; molecules that do so are hydrophobic. Grease and fats are hydrophobic, whereas salt is hydrophilic ("water-loving")

ICE BEER Beer produced with a process including ice generation

IMMOBILIZED YEAST Yeast attached to an insoluble support (e.g., glass beads) that can be used in continuous processing whereby wort or beer is flowed past it

INDIRECT HEATING   Heating of a material without direct application of heat, but rather via a heat exchanger

INFESTATION   Condition whereby a raw material in the malting or brewery has animal life within it, for example insects in badly stored barley

INORGANIC   Any chemical species other than those containing carbon (carbon dioxide, despite containing carbon, is regarded as inorganic)

INSECTICIDE   Material sprayed onto crops either during growth or storage to eliminate insect infestation

INVISIBLE HAZE   Haze that registers on a hazemeter but is not perceptible to the eye; sometimes called pseudohaze

IRON   Inorganic element that can enter into beer from some raw materials (e.g., filter aids) and potentiate oxidative damage

ISINGLASS   Preparation of solubilized collagen from the swim bladders of certain fish, used for clarifying beer; normally referred to as "finings"

ISO-α-ACID   Bitter component of beer derived from hops

ISOMERIZATION   The conversion of hop α-acids into iso-α-acids, achieved during wort boiling

KEG   Large container for holding beer, for subsequent draft dispense by pump

KETTLE   Brewhouse vessel in which wort is boiled; also known as "copper"

KIESELGUHR   Mined powder, derived from skeletons of microscopic animals, used to aid the filtering of beer

KILNING   Heating of germinated barley to drive off moisture and introduce desired color and flavor

KRAUSENING   Traditional German fermentation practice in which fresh fermenting wort is introduced late during warm conditioning to stimulate maturation of beer

LACING   Tendency of beer foam to stick to the side of the glass (also known as cling)

LAGER   A type of beer, traditionally pale, produced by bottom-fermenting yeast and produced in a relatively slow process, which includes lengthy cold storage ("lagering"). The word *lager* is derived from the German *Lager* "storehouse"

LARGE PACK   Kegs or casks

LATE HOPPING   Practice of adding a proportion of the hops very late in the wort boiling phase in order to retain certain hop aromas in the ensuing beer

LATE HOP ESSENCES   Extracts that can be added to beer to introduce a late hop character

LAUTER   The act of separating sweet wort from spent grains; also, the vessel used to perform this duty

LEAD CONDUCTANCE VALUE   A method for assessing how much bitterness precursor is present in hops

LIGHT (LITE) BEER   Beer in which a greater proportion of the sugar has been converted into alcohol

LIGHTSTRUCK   Skunky flavor that develops in beer exposed to light

LIMIT DEXTRINASE   Enzyme in malt that breaks the branch points in the amylopectin component of starch

LIPID   A material that does not dissolve in water, but does dissolve in organic solvents

LIQUID CARBON DIOXIDE   Solvent produced by liquefying carbon dioxide gas at low temperatures and high pressures; used for extracting materials from hops

LIQUOR   Water

LOW-ALCOHOL BEERS   Beers containing a low level of alcohol (e.g., less than 2% ABV, although the definition differs between countries)

LUPULIN   The glands in hop cones that contain the resins

MALT   Dried germinated barley

MALTING   The controlled germination of barley involving steeping, germination, and kilning so as to soften the grain for milling, develop enzymes for breaking down starch in mashing, and introduce color and desirable flavors

MALTING GRADE   Score allocated to a barley variety that indicates whether it will give a high hot water extract after conventional malting and mashing

MASHER   Device positioned before the mash mixer that facilitates intimate mixing of milled malt and hot liquor

MASHING   Process of contacting milled grist and hot water to effect the breakdown of starch (and other materials from the grist)

MASH FILTER   Device incorporating membranes for separating wort from spent grains

MASH TUN   Vessel for holding a "porridge" of milled grist and hot water to achieve conversion of starch into fermentable sugars

MASHING OFF   Conclusion of mashing, when the temperature is raised prior to the wort separation stage

MATURATION   The post-fermentation stages in brewing when beer is prepared ready for filtration

MEALY   Favorable texture of the starchy endosperm of barley that makes it easy to modify

MELANOIDINS   Color contributors in beer produced by the reaction of sugars with amino acids during heating stages in malting and brewing

MEMBRANE   A sheet, either one found in a living system (e.g., the plasma membrane that surrounds a yeast cell) or one that has a specific job to do in a brewery (e.g., in a mash filter or a beer filter)

METABOLISM   The sum of the many chemical reactions involved in the life of an organism such as barley or yeast

MICROPYLE   The area at the embryo end of a barleycorn through which water can gain access

MILLING   The grinding of malt and solid adjuncts to generate particles that can be readily broken down during mashing

MITOCHONDRION   The organelle in a eukaryotic cell responsible for generating energy in respiration

MODIFICATION   The progressive degradation of the cell structure in the starchy endosperm of barley

MOISTURE CONTENT   The amount of water associated with a material such as barley, malt, hops, or yeast

MOLD   Infection of barley or hops

MOUTHFEEL   The "tactile" sensation that a beer creates in the mouth (also referred to as texture)

NEAR-INFRARED A region of the light spectrum where wavelengths are longer than those in the visible red region, but shorter than those in the infrared region; NIR spectrometers are increasingly widely used for making various rapid measurements in the malting and brewery

NITROGEN There are two completely separate meanings for nitrogen in malting and brewing: (a) the nitrogen atom as it is found in proteins; its level in barley, malt, or wort is a measure of how much protein they contain; (b) gaseous nitrogen (N_2), which is sometimes introduced into beer to enhance foam. This process of introduction is called nitrogenation

NONALCOHOLIC BEERS Beers containing very low levels of alcohol, for example less than 0.05% ABV (although the definition differs between countries)

NONENAL Compound that contributes to the cardboard character that develops in stale beer

NONRETURNABLE BOTTLES Glass bottles that are not returned to the brewery for refilling; also referred to as "one-trip bottles"

NUCLEATION Spontaneous bubble formation in a wort or beer

NUCLEIC ACIDS The complex polymeric molecules in living systems that are responsible for carrying the genetic message and translating it

ORGANELLE A distinct region within a eukaryotic cell with its own specific function

ORGANIC Refers to compounds containing carbon (apart from carbon dioxide)

ORGANIC ACIDS Carbon-containing acids such as citric and acetic acid released by yeast, largely responsible for the pH fall during fermentation

ORGANOLEPTIC Pertaining to taste and smell

ORIGINAL EXTRACT The amount of extract present in a starting wort as calculated from the amount of nonfermented extract left in a beer together with the amount of extract equivalent to the amount of alcohol produced in a beer. In some countries this is known as original gravity

OSMOTIC PRESSURE The force that drives water to pass from a dilute solution to a more concentrated one through a semipermeable membrane

OXALIC ACID An organic acid found in malt that must be precipitated out in the brewhouse by reacting with calcium to form the calcium salt. Otherwise it will precipitate out in beer as "stone"

OXIDATION At its simplest, process of deterioration of beer due to ingress of oxygen

pH A measure of the acidity/alkalinity of a solution

PALE ALE English-style ale, usually in small pack

PAPAIN Protein-hydrolyzing enzyme from papayas

PASTEURIZATION Heat treatment to eliminate microorganisms

PENTOSAN Polysaccharide located in cell walls of barley

PEPTIDE Molecule consisting of perhaps 2–10 amino acids linked together

PERLITE Volcanic ash used in the filtration of beer as an alternative to kieselguhr

PESTICIDES   Agents used to protect crops from infection and infestation during growth and storage

PIECE   The bed of grain in a malting

PILSNER (OR PILS)   A style of lager originating in the Czech Republic

PINT   A measure of beer volume (473 ml in the United States; 568 ml in the United Kingdom)

PITCHING   The introduction of yeast into wort prior to fermentation

PLATE-AND-FRAME   A type of beer filter

PLATO   Unit of wort strength

POLYPEPTIDE   A partial breakdown product of proteins containing approximately 10–100 linked amino acid units

POLYPHENOL   Organic substance originating in husk of barley and also in hops that can react with proteins to make them insoluble; also known as tannin

POLYSACCHARIDE   Polymer comprising sugar molecules linked together

POLYVINYLPOLYPYRROLIDONE (PVPP)   Agent capable of specifically binding polyphenols and removing them from beer

PRE-ISOMERIZED EXTRACTS   Extracts of hops in which the α-acids have been isomerized

PRIMINGS   Sugar preparations added to beer to sweeten it

PROPAGATION   Culturing of yeast from a few cells to the large quantities needed to pitch a fermentation

PROPYLENE GLYCOL ALGINATE   Material added to beer to protect the foam from damage by lipids

PROTEASE   Enzyme that breaks down proteins

PROTEIN   Polymer comprising amino acid units

PROTEOLYSIS   The breakdown of proteins by proteases

PSEUDOHAZE   Invisible haze

PURGING   Elimination of an unwanted volatile material (e.g., a flavor or a high CO_2 or O_2 content) by bubbling through N_2.

QUALITY ASSURANCE   Approach to quality maintenance that involves establishing robust processes and systems designed to yield high-quality product

QUALITY CONTROL   Monitoring of a process to generate information that is used to adjust the process in order to ensure the correct product

RACKING   The packaging of beer

REAL EXTRACT   The amount of dissolved material in beer that has not been converted into alcohol

REDUCED HOP EXTRACTS   Pre-isomerized extracts that have been chemically reduced so that they are no longer light-sensitive and can be used to provide bitterness to beers that are intended for packaging in green or clear glass

REFRACTOMETER   Device for measuring the strength of beer

REPITCHING   Practice of taking yeast grown in one fermentation to pitch the next batch of wort

RESIN   Substances from hops that generate the bitterness in beer

ROUGH BEER   Beer before filtration

SACCHAROMYCES CEREVISIAE   Brewer's yeast

SALADIN BOX   Type of vessel for germinating barley

SCREENING   Cleaning of unwanted debris from barley

SEAM   The "join" between a beer can and its lid

SMALL PACK   Cans and bottles

SOLUBLE NITROGEN RATIO   The ratio of the dissolved nitrogen (protein) in wort to the total nitrogen (protein) in malt, which is in direct proportion to the nitrogen modification (sometimes called the Kolbach index)

SPARGING   Spraying the spent grains with hot water during the wort separation process to facilitate extraction of dissolved substances

SPECIFICATION   A parameter measured on a raw material of brewing, on a process stream, or on the finished beer, which must be within defined limits for the material to pass to the next stage in the process

SPECIFIC GRAVITY   The weight of a liquid relative to the weight of an equivalent volume of pure water (also referred to as relative density)

SPECTROPHOTOMETER   Device for measuring the amount of light absorbed by a solution

SPENT GRAINS   The solid remains from a mash

SPOILAGE ORGANISM   Microbe capable of infecting wort or beer

SQUARE   Style of fermenter in that shape

STABILIZATION   Treatment of beer in order to extend its shelf life

STALING   Deterioration in the flavor of beer

STARCH   Polysaccharide food reserve in barley

STEELY   Texture of starchy endosperms in those barleys that are less suited to producing good quality malts for brewing

STEEPING   Increasing the water content of barley by soaking

SUGAR   Small, sweet carbohydrate

SULFUR COMPOUND   Flavor-active material in beer containing sulfur atom(s)

SWEET WORT   Wort prior to boiling with hops

SYRUP   Concentrated solution of sugars

TAINT   Off flavor in beer or a raw material

TANNIC ACID   Material added to beer to precipitate out protein

TETRAZOLIUM   Dye used to detect whether barley is alive

THREE-GLASS TEST   Procedure for blind tasting to discern whether two samples of beer can be differentiated

TINTOMETER   Device consisting of a series of colored discs for comparing with a beer to ascertain whether it has the correct color

TOP FERMENTATION   Fermentations in which the yeast collects at the top of the vessel

TOTAL SOLUBLE NITROGEN   A measure of the dissolved protein in wort

TRIGEMINAL SENSE   Sensation of pain detected by the trigeminal nerve

TRUB   Insoluble material emerging from wort on heating and cooling

TUNNEL PASTEURIZATION   Pasteurization of small pack beer by passage through a heated chamber

ULTRAFILTRATION   Filtering out of material at the molecular level by passage through very fine membranes

VIABILITY   Measure of how alive something is

VICINAL DIKETONES   Butterscotch-
flavored compounds formed during
brewery fermentation

VIGOR   A measure of the strength of
growth of the barley embryo during
germination

VISCOSITY   A measure of how much a
liquid resists flow

VITALITY   A measure of the healthiness
of a living yeast

VOLATILE   A molecule in beer that contributes to aroma and is easily driven off

VORLAUF   Recycling of the first wort
runnings from a lauter tun in order to
ensure "bright" wort

WATER SENSITIVITY   Tendency of a
barley's germination to be suppressed
by the presence of excess water

WEAK WORT RECYCLING   Use of the
weaker worts from the end of wort
separation to mash in the next mash

WHIRLPOOL   Vessel for separating trub
from boiled wort

WORT   Fermentation feedstock
produced in the brewery

WORT SEPARATION   Act of separating
sweet wort from spent grains

XEROGEL   Colloidal stabilizing agent
(similar to hydrogel) made from silica

YEAST   Living eukaryotic organism
capable of alcoholic fermentation

YEAST FOOD   Source of amino acids
and vitamins sometimes used in
brewery fermentations

ZENTNER   Unit of hop mass
(50 kilograms)

FURTHER READING

This book is not filled with references. This has been a deliberate policy, for in most instances the most relevant citations would be to somewhat erudite scientific and technical journals, written for the specialist and unlikely to be readily digested by the layperson.

Indeed, most of the books covering the brewing process are somewhat technical, but I am able to recommend some volumes that will appeal variously to those with different extents of scientific education.

For a basic appreciation of brewing from a more molecular perspective than I have offered in this book, I have no hesitation in steering you toward my own *Scientific Principles of Malting and Brewing* (American Society of Brewing Chemists, 2006).

For those in search of much more detail on barley, the volume of choice should be *Barley: Chemistry and Technology,* edited by A. W. MacGregor and R. S. Bhatty (American Association of Cereal Chemists, 1993), while for hops go to *Hops* by R. A. Neve (Chapman & Hall, 1991). Yeast is covered extensively by two of my dearest pals in the industry, Chris Boulton and David Quain, *Brewing Yeast and Fermentation* (Blackwell Science, 2001), while two more friends address the beer itself in *Beer: Quality, Safety and Nutritional Aspects* by E. Denise Baxter and Paul S. Hughes (Royal Society of Chemistry, 2001). In terms of quality measurements, it would be strange of me not to recommend another of my own offerings, *Standards of Brewing: Formulas for Consistency and Excellence* (Brewers Publications, 2002). In for a penny... the reader interested in a relatively impartial comparison of beer and wine for their history, sociology, scientific complexity, technical excellence, variety, suitability for pairing with food, and health-giving properties might like to take in my *Grape versus Grain* (Cambridge University Press, 2008).

For a voluminous read of the proud history of beer, then go no further than Ian Hornsey's *A History of Beer and Brewing* (Royal Society of Chemistry, 2006).

INDEX